EXCEL® MANUAL

MARK DUMMELDINGER
University of South Florida

STATISTICS FOR BUSINESS AND ECONOMICS

ELEVENTH EDITION

James T. McClave
Info Tech, Inc
University of Florida

P. George Benson
College of Charleston

Terry Sincich
University of South Florida

Prentice Hall
is an imprint of

Reproduced by Pearson Prentice Hall from electronic files supplied by the author.

Publishing as Prentice Hall, 75 Arlington Street, Boston, MA 02116.

ISBN-13: 978-0-321-64180-9
ISBN-10: 0-321-64180-9

1 2 3 4 5 6 BRR 13 12 10 09

Prentice Hall
is an imprint of

www.pearsonhighered.com

PREFACE

Excel is an extremely powerful and useful software program with many different capabilities. Statistical analysis is just one of those capabilities. Many of the tools and techniques that you will be learning in your business statistics course can be calculated within Excel. Sometimes, however, a technique is not provided within the standard Excel package or it can be difficult to conduct within Excel. That is where the DDXL add-in can help.

DDXL offers very easy-to-use menus to conduct the majority of the analyses used in today's business environments. This manual is intended to walk you step-by-step through the analyses that you will be learning about in class. Most of the instruction in this manual focuses on the DDXL add-in program. There are several instances, however, where Excel functions and data analysis techniques are demonstrated. When both DDXL and Excel can be used, I show you the DDXL method as I feel it is generally the easier to work with.

This manual has been organized to present techniques to the reader in the context of examples. I have used the example data sets presented in the text so that the reader can conduct the analyses desired and compare the results and output to those presented in the text. At the end of most chapters, a Technology Lab presents additional homework exercises to the reader to practice on their own. I have also included the final output so that the reader can verify that they are correctly conducting the analyses that they desire.

A NOTE ON DATA SETS

Many example and all exercise data sets referenced in this manual are available on the CD that is bound inside the textbook. However, there are a few example data sets that are not available on this CD. To access these files (along with all other example data sets used in this manual), please visit the following Pearson Datasets website:

http://media.pearsoncmg.com/ph/esm/statistics_datasets/stats_datasets.html

Click on the link for McClave, Benson & Sincich and follow that link to the eleventh edition of the text. Here, you will find all of the text's datasets along with this Excel manual's data sets. You should be able to download the Excel data sets from this site to your computer. Simply open these files in Excel and follow the directions given in this manual.

My goal was to write a manual that helps as you learn the different tools and techniques presented in your introductory statistics course. I hope that this manual can alleviate some of the fear and frustration of the numerical calculations that are inherent to any statistical analysis. Ultimately, I hope that you begin to see and master the power of the information that can be learned through the techniques I illustrate in this manual. Good luck!

Mark Dummeldinger

Table of Contents

Primer
Excel Basics Needed for Statistical Analysis of Data

P.1 Introduction and Overview

This manual is designed for use with McClave, Benson, and Sincich's *Statistics for Business and Economics*, 11th Edition. It is not intended to take the place of your Excel user's manual, however it will introduce the Excel novice to the software and provide the basic tools necessary to analyze statistical data using Excel. To accomplish this goal, we follow a four-part process. First, we introduce the statistical procedures available in Excel. Next, we illustrate these procedures by teaching you how to perform the Excel commands required to produce the output from selected examples in *Statistics for Business and Economics 11/e*. Both the steps taken and output generated are provided in this manual to teach you the Excel steps to be followed. Third, we have provided an Excel Lab where the student is given Excel workbooks to use in answering step-by-step questions. The final part of our process is to provide you with Excel data sets that may be used to complete homework exercises in the McClave/Benson/Sincich text. Our hope is that this "introduce-learn-practice" format will enable you to finish the course with a firm understanding of how Excel can be used to analyze statistical data.

We believe that the Excel portion of a statistics course should serve strictly to enhance the statistics that is being taught. We have tried to keep this philosophy in mind when writing this manual. We have attempted to provide an easy-to-use format that will allow you to use Excel to calculate the statistics you learn in class. If we have been successful, you will view Excel as a valuable tool for the statistician. Used correctly, Excel allows the statistician to spend more time using, and less time calculating, the kind of information that you will explore in your statistics course.

P.1.1 Versions of Excel

This manual principally uses the 2007 version of Microsoft Office Excel software available from Microsoft. Many different versions of Excel exist, however all of the versions function in essentially the same way. You will notice slight differences in how the screens look and in the names of some commands. In this manual, we will primarily use the statistics procedures available in the DDXL add-in currently available for Excel. When possible, we have supplemented these procedures with different Functions and techniques available within the Excel program. Our goal throughout this manual is to provide the user with the easiest method of generating the desired output.

P.1.2 Versions of Windows

The copies of screens shown in this book are taken from a PC using Windows Vista. They will appear slightly different if using another operating system or a Macintosh computer. After you are operating comfortably within Excel, these differences should be minor. There will, however, be slight differences between Macs and PCs in the keys used for commands.

P.1.3 What May Be Skipped

If you have used spreadsheets before, you can probably omit much of the first chapter. Other programs, such as Lotus 1-2-3 and Quattro, use slightly different terminology when describing the spreadsheets. The concepts, however, are essentially the same although the terms and/or procedures may differ slightly.

- For those of you who are looking for commands for a specific procedure or for an example from *Statistics for Business and Economics*, 11/e you will find that each of the chapters may be used independently.

P.1.4 More Detailed Information on Excel

A glance at the bookshelves in the computer section of most bookstores will reveal a number of books that deal with Excel in all its various versions. Few deal with Excel as a way to perform statistical analyses. Use care when selecting these resources to help with statistical analyses of data.

P.2 What You Need to Know to Begin Using Excel

P.2.1 Using the Mouse

Mice come in several forms. The majority are provided with new computers and roll on the desktop or pad. A small ball on the bottom, when rolled, causes the pointer on the screen (called the **screen pointer**) to move in a corresponding way. Another version (called a trackball) places a larger ball in a framework that allows you to roll the ball with your fingers. Finally, there are other forms that have small screens that you move your fingers across as you would move the mouse. The pressure of your finger moving across the screen causes a screen pointer to move in synchrony with your movements.

All devices have at least one, and most likely two or more, buttons that you can click or hold down, sometimes while also moving the ball. There are four basic actions you will need to use in operating the mouse:

- **Point** - You point to objects on the screen by sliding the mouse on the desk pad or rolling the trackball. The screen pointer will track the movements made on your desk. The shape of the screen pointer will change, most often being an outline arrow or the outline of a plus sign when using Excel, but changing with the task to be done.

- **Click** - "Click" means to press and release the left mouse button (called a left-click). If you are pointing at an executable command, this action will cause it to take place. If you point to any cell on the spreadsheet and click, that cell becomes the **active cell** and is ready to receive data. Sometimes you may be asked to press and release the right mouse button (called a right-click), which is commonly used to place a shortcut menu on the screen. The above assumes you are using the settings for the mouse provided by the manufacturer with your right hand. If you are left-handed or want to reverse the way the buttons function, this can be done. Click on the **Help** icon within your version of Windows and look for *mouse*, *buttons*, *reversing* in the **Index**.

- **Double-click** - "Double-click" means to press and release the left mouse button twice rapidly. If you fail to press rapidly enough, it is interpreted as one click. Often this process replaces the two-step sequence of selecting a command and then clicking on **OK** to execute that command.

- **Drag - Objects** on the screen are moved by dragging. To drag, place the mouse pointer on the item you want to move, click and hold the left mouse button --- do NOT release it. While you hold the mouse button down, slide the mouse to move the screen pointer and the item to the location you want. Then release the mouse button.

P.2.1.1 Starting and Exiting DDXL in Excel

To start DDXL in Excel 2007:

Click on **Add-Ins** tab at the top of the screen Excel workbook. Move the mouse and click on the **DDXL** menu located in the upper left-hand portion of the workbook. Use the pull-down menu to select the analysis that you would like to use.

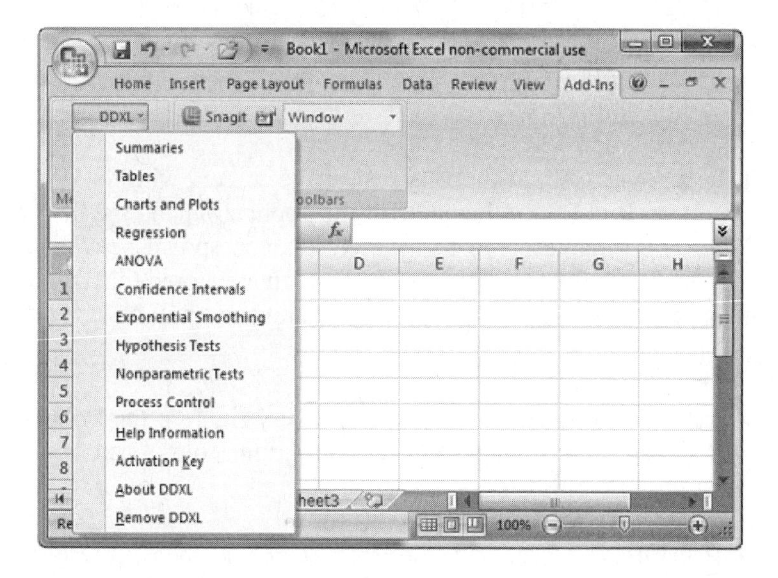

To end DDXL in Excel:

Simply finish the analysis you are conducting and click on the

X located in the upper right-hand corner of the DDXL Viewer.

P.2.1.2 Starting and Exiting Excel

To start the Excel program:

Click on **Start** in the lower left of the screen. Move the mouse to **Programs** and then continue moving through menus until you find the **Microsoft Excel** icon to click and begin.

If you have an Office Suite of programs, you may have to find the folder containing the suite, open it, and double-click on the icon of the Excel program you find there.

You may want to go through the Quick Preview online tutorial if you are unfamiliar with spreadsheets and need a quick overview. This is often found in the Excel folder.

To exit the program using either Windows-based version:

> Click in the upper **X** (the **close button**) that you find in the upper right corner of the screen. (If two sets of boxes are showing, the lower set applies to the worksheet (spreadsheet) you have showing while the upper one is for the application or program itself, (i.e., Excel)). If you have edited (changed) any of the information in the **workbook**, you will be prompted to save the information before closing the program. In many computer labs, you may be asked to save all of your data on a diskette. We will describe this procedure later.

P.2.2 Layout of Worksheets and Workbooks

The figure below shows what is on the screen when Excel is opened. It may be slightly different from what you have on your screen. The list that follows briefly describes some of the items located in the Excel program that the figure above, starting at the top.

- **Program Title Bar** - This is most likely at the very top of the screen, with the default title being Book 1 - Microsoft Excel. It indicates the name of the application and the fact that you are in what is called *Book 1*, the name given to a newly opened spreadsheet. Each book initially consists of three worksheets (this can be changed) which are stored together as a unit called a book. When you save your work, all of the sheets in this book will be saved together as one file.

- **Program Icons** - Known as **sizing buttons**, these are at the very right on the Program Title Bar, just as on the screen shown above. There are three program icons. Each is described below.

  - The **minimize button** shrinks the program which is then represented as a button on the taskbar at the bottom of the screen. This is a part of Windows, which inactivates, but does not close, programs and places their icon at the bottom of the screen so they can be immediately reactivated.

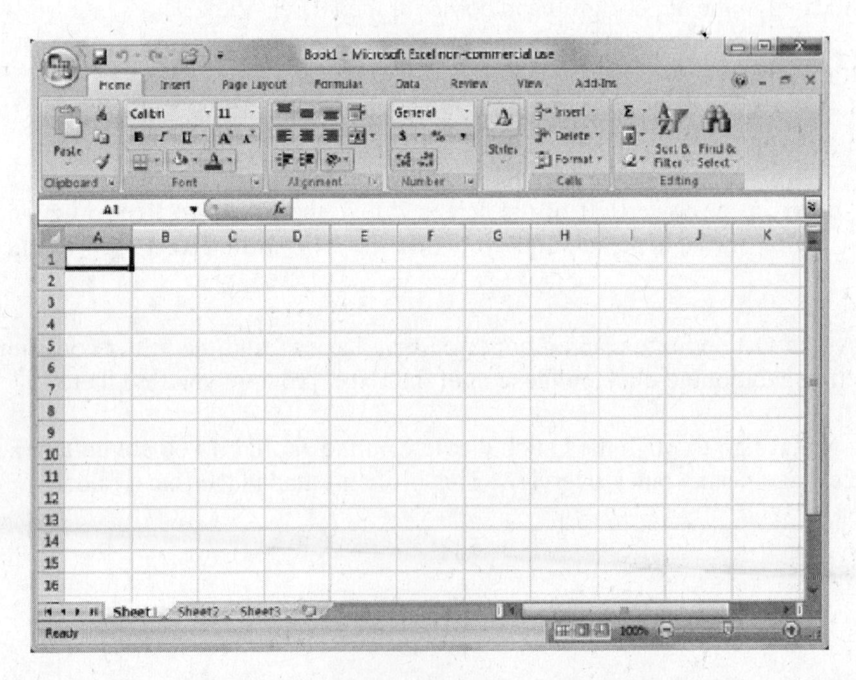

- and . The **maximize and minimize buttons** switch between the full-screen and window views.

- The **close button,** which closes the Excel application. You will be asked if you want to save your work, if there is any, before the application is closed.

- The **Ribbon** – Applications and commands are grouped together by task into the eight tabs on the ribbon. In the figure below, the Home tab has been selected. Each tab is divided into different groups that contain features within Excel that are somehow related to each other. Those that are useful for our purposes will be discussed later.

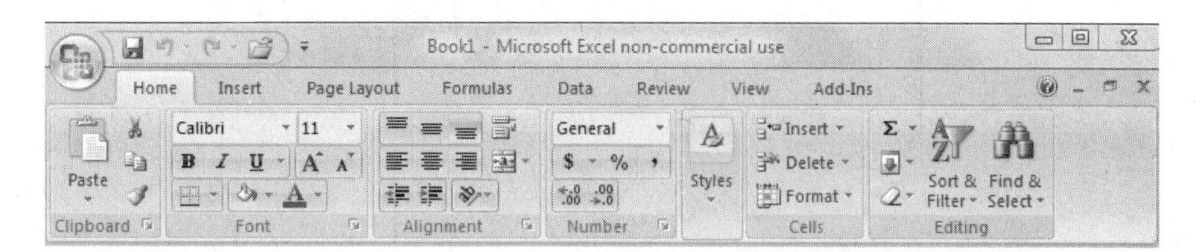

The **Microsoft Office Button** – Located in the upper left hand corner of the screen, this button contains basic file operations**.** This button is shown in the figure to the right.

- The **Quick Access Toolbar** – Located next to the Microsoft Office Button in the figure above, this toolbar offers quick access to various commands and actions within Excel. By clicking on the , this toolbar can be customized.

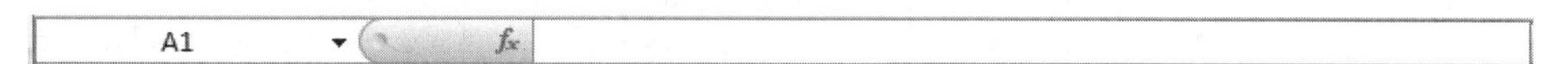

- The **Formula Bar -** Located just below the Ribbon in the worksheet, the Formula Bar, shown above, consists of two windows. The first window, called the Name Box, shows the address of the active cell in the worksheet that is displayed. Initially it is A1, so this is the address displayed. Move to a different cell and click on it to make it the active cell. Now that cell's address is displayed.

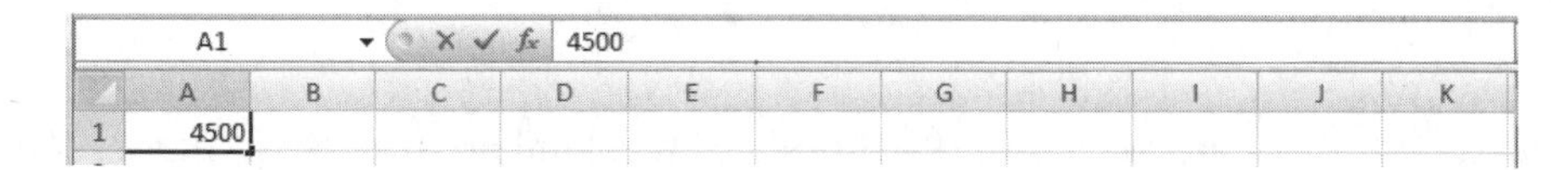

Notice what happens to the Formula Bar when we type some numbers in cell A1. Three buttons appear:

- The is clicked when we want to "destroy" or delete the information we have typed in the active cell. That information is also displayed in the Formula Bar and will be removed from both sites is we click on the X.

- The , when clicked, indicates that the data as entered are acceptable. The data will remain in the cell, but the three buttons disappear indicating that the cell is not being edited.

- Finally, the f_x is an icon to turn on the **Function Wizard**, a set of over 300 functions in categories such as Financial, Math & Trig, Statistical, etc. We will use the statistical functions quite often, and discuss the use of this tool throughout the chapters.

- **Worksheet Area**. Finally, we have the worksheet, which consists of cells with columns labeled as letters and rows as numbers. Each cell is identified by the combination of its column letter and row number as is displayed in the Name Box.

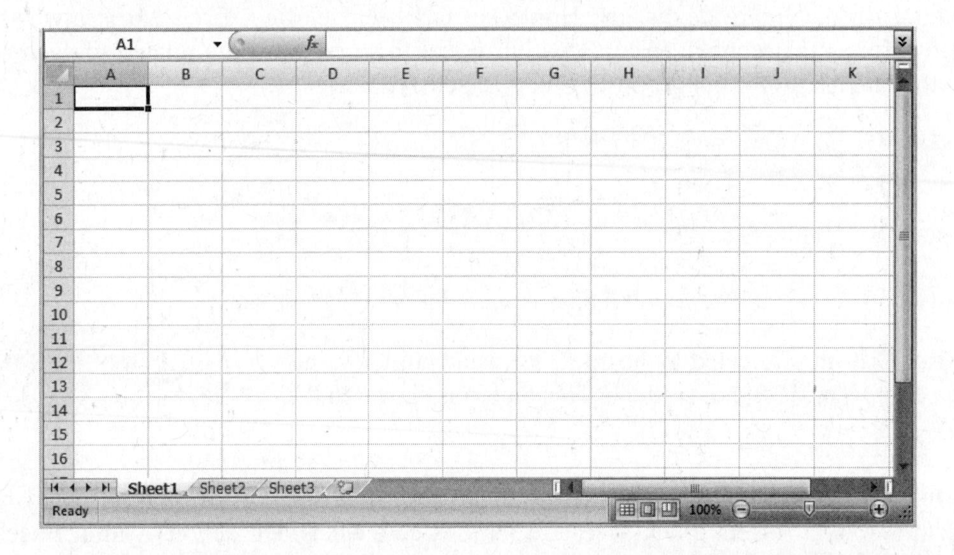

- **Scroll Bars** are found at the right and at the bottom of the worksheet. There are two small arrows, which look like a triangle laid on its side at either end of each scrollbar that, when clicked with the mouse, will move the screen up, down, right, or left one line for each press. Some users call the box within the scroll bar an *elevator*. You can grab the box or elevator and drag it. The screen will move a distance that corresponds to the amount you move the box. Finally, if you click in the shaded area between the box and the top or bottom of the scroll bar, the screen will move one whole screen in the direction you click. Thus, you can move a line at a time, a screen at a time, or from the top to the bottom of the screen. Also, if you hold down the shift key while dragging the box, you can move to the end of the row or column.

- **Worksheet Tabs**. These are at the lower left of the screen and are labeled as *Sheet 1*, *Sheet 2*, etc. The default setting provides three of these tabs, which you can move through by clicking the arrows to the left of the name. Try it. Two arrows will move the active sheet to the left: one moves a sheet at a time and the other will move to the left-most sheet. The same applies to the right arrows. All three of these worksheets are stored together as one unit, called a Book.

P.3 Ways to Get Help

P.3.1 Help on the Main Menu

The Help Button, , is located on the right-hand side of the Ribbon. Click on it and you will see the Excel Help menu shown below.

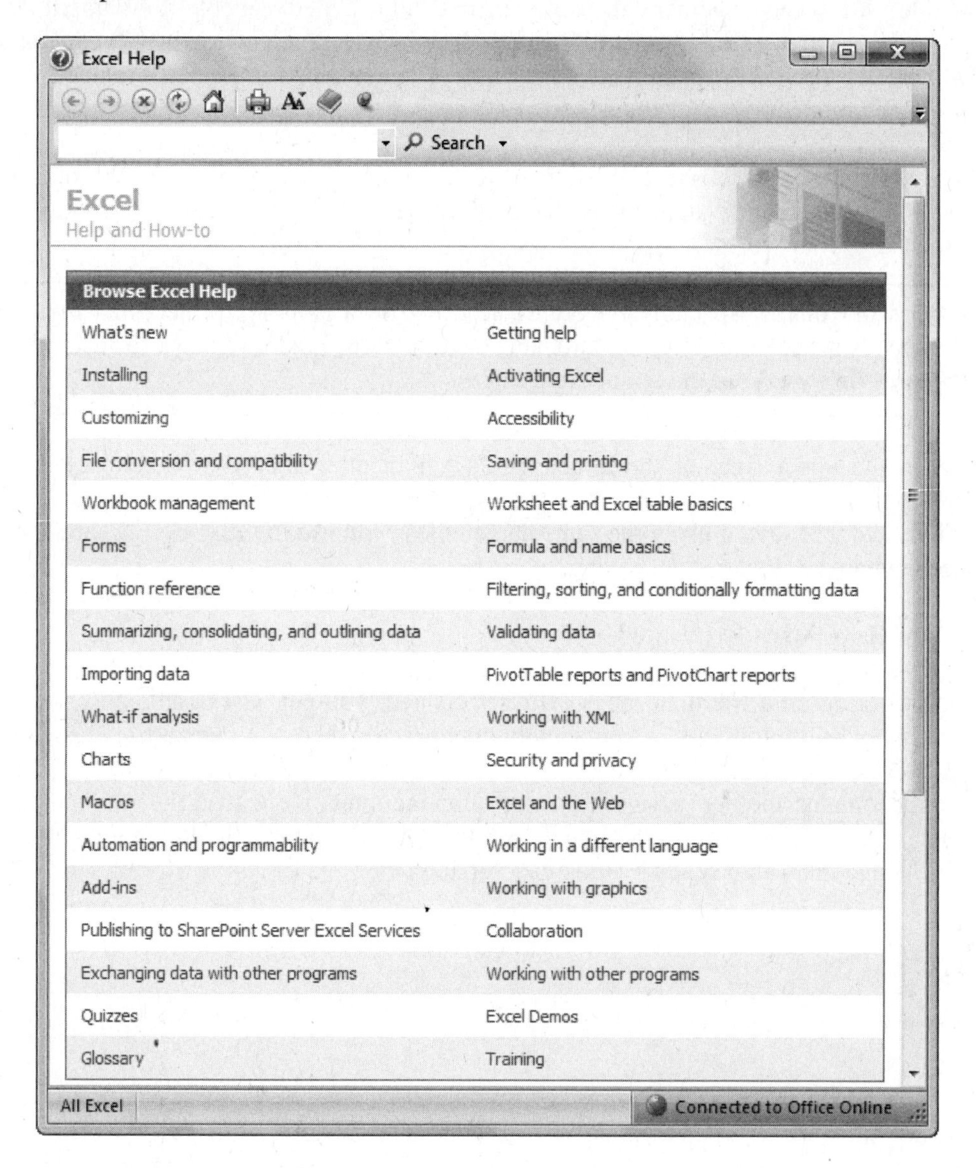

Within this menu, there are many topics to search under to get help within Excel. Or you can search for a topic in the Search box at the top of this menu.

P.4 Opening and Saving Documents

P.4.1 Opening a Brand New Spreadsheet

If you started Excel by clicking the icon, the screen opened with a blank spreadsheet. The title at the top indicates it is called *Book 1* and the tabs across the bottom are labeled *Sheet 1, Sheet 2*, etc. Initially there are three sheets available. All sheets together are stored in one unit, called a book. You might, for example, choose to keep all homework assignments for one class in one unit, now called *Book 1*, but renamed by you as *Stats 1 HW*. You could then create other books that contain your personal budget, your records as the treasurer of an organization, and so on.

When you enter information into the worksheet, it is stored in the active memory of the computer, and will disappear is power is lost, whether by an accident that deprives you of electrical power or by turning off the machine without saving your work.

> It is wise to save your work often, especially if you are working on a complex project that requires many hours of data entry or processing. Use the automatic save feature. Find directions for using it in **saving documents**, **protecting work**, in the list of **Help** topics!

If you click on that X in the upper right of the Excel window, indicating that you want to quit, the program will ask you if you want to save the file you have created. If you have already saved it, and therefore have named it, Excel will save it using the same name unless you use the Save As … command, which is used to change the name or location for saving.

P.4.2 Opening a File You Have Already Created

If you want to continue working on a file that you previously created, you can just double-click on the icon that represents that file. The program will automatically open Excel when you open a file created by Excel. Sometimes this causes a problem if you create a file in one version of Excel and want to open it using a computer that only contains another version. Files are often incompatible with earlier versions and cannot be opened by an earlier version. One solution is to use **Save As…** and save the file you created in the newer version in the format that can be used by the older version.

P.5 Entering Information

In the previous chapter we described the different parts of the Excel screen – the Ribbon, the Microsoft Office Button, the Program Title Bar, Program Icons, the Quick Access Toolbar, the Formula Bar, Worksheet Areas, Scroll Bars, and Worksheet Tabs. Now we are ready to enter, save, edit, retrieve, and perform other data manipulations.

- **Addresses** Each cell is identified by a combination of a letter and a number to locate it in the spreadsheet. Letters for each column are shown across the top of the worksheet and begin to repeat with two and three letter combinations after Z is reached.

P.5.1 Activating a Cell or Range of Cells

When the spreadsheet is initially opened, Cell A1 is automatically the active cell. It has a dark outline around it, which indicates that whatever you type will be entered into that cell. Note that the address A1 is displayed in the Name Box, which is to the left of the Formula Bar. Move to another cell, click, and note the change in the Name Box.

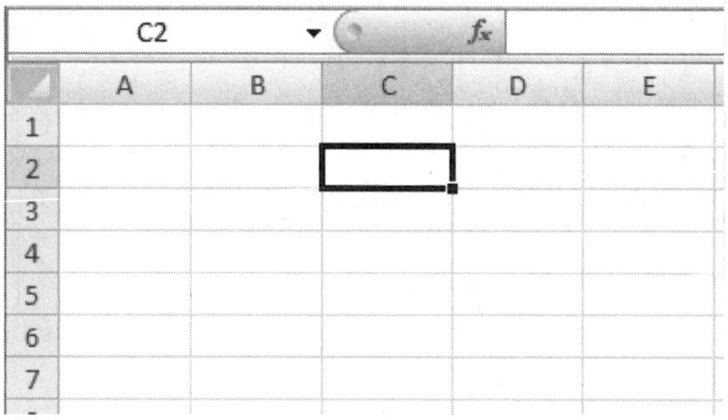

Often we need to refer to more than one cell at a time. A group of cells is called a **Range**. Click on cell B4, hold the mouse button down and drag down to B8. Release the button. The screen is now darkened (highlighted) in the range of cells, except for the top cell, B4. To indicate the address of a range of cells, we separate the addresses of the upper left and lower right cells with a colon. Here we have a range identified as B4:B8, although the Name Box only indicates the address of the top cell.

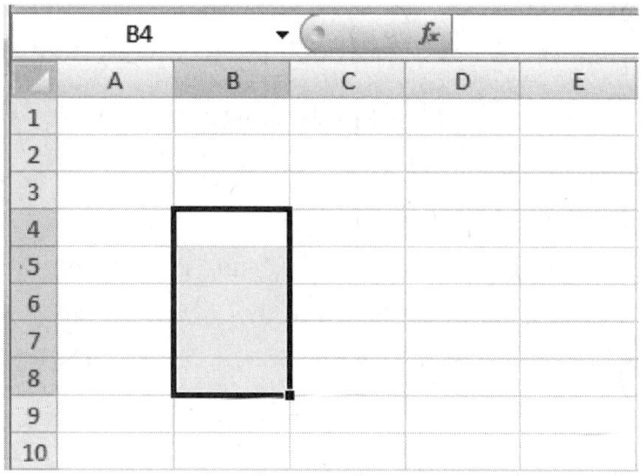

To activate cells in many rows and columns (i.e., a range) place the cursor in the upper left cell and drag to the lower right. Now all of the cells in that range will appear shaded. You can click and drag in the opposite direction if you wish. This is handier if you tend to overshoot your target and continue on past where you want to end, as many of us do. Another way to activate a range of cells is to click in the upper left cell, move to the lower right cell using the scroll bars or arrows, and then press the SHIFT key and click on the lower right cell (SHIFT + CLICK ON CELL) at the same time. Now all cells between these two points are shaded.

P.5.2 Types of Information

There are three types of information that you can enter into a cell:

1. Text. This is the term used by Excel developers. Other spreadsheet programs may call the alphabetic characters typed in a cell "labels.
2. Numbers. Most often you will enter numerical data.
3. Formulas. These cause new information generated from operations performed on text and numbers that are entered in cells.

P.5.3 Changing Information

To change information in a cell, you have to consider which of two situations exist:

- If you have not yet "accepted" the information by clicking the green check mark (or pressing enter, or using arrow keys, or ...) then you can simply use the backspace or delete keys to remove entries. Insert the **I-beam** at the point where you want to change something and use the backspace key to remove characters to its left and the delete key to remove them at its right.

- If you are typing information into a cell and decide you want to start over, click on the red X and everything will be deleted.

- If you just want to delete everything in the **active cell** or range of cells, press the delete button.

- If you have already entered and accepted data in a cell, but now want to go back and edit it but not erase all of it, activate the cell and then insert the I-beam in the editing window where you want to make the changes.

As a practice exercise for changing information, try going through the following steps:

- In cell A1 type 12346. Press Enter, which moves you to A2.
- Assume you really wanted to enter 123456.
- Return to A1 by using the arrow key or mouse.
- Move the mouse pointer so the I-beam is between 4 and 6 in the editing bar. Click once.
- Note that the three editing keys are now shown to the left of 12346.
- Type 5, which will be inserted between 4 and 6

A1 fx 123456

	A	B	C	D
1	123456			
2				
3				
4				

P.5.4 Moving and Copying Information

A basic principle used in many programs is that you mark or indicate which material will have something done to it by first marking it and then execute the command that does something to the highlighted material. We will see this principle operate in several other places in the program. When we want to move information we can do it so that it is removed from one location and placed in another. This is a **cut**. If we want to make a copy of the material so that it is in the original location as well as in other locations we make a **copy**. Help for each of these procedures is obtained by searching the Help menu using the words *cutting* or *copying*.

To copy the entry in cell A1 (123456) to cell A2, we first mark the cell by clicking on it. Move the mouse pointer to the icon for copy, , found in the Clipboard in the home tab on the Ribbon. The outline around the cell will become like the lights on a marquee; it will alternately darken and lighten. It is ready to be copied. Click the cell where you want the copy, say A2. Click on the icon for paste to place a copy there. Copy operates like a rubber stamp; we have a copy stored in the Clipboard and can continue to place copies anywhere we wish. Activate another cell, say C3 and paste another copy there.

Cutting is done in a similar fashion, except that the cell is empty after you cut the data from it. Click on any cell with content, say A1, and then click on the cut icon, , in the standard toolbar, which is a pair of scissors. The cell will be outlined, as it was when you used *copy*. Activate another cell and click on the icon for pasting, the same one used before.

If you cut or copy a group of cells, i.e., a range, the principle is the same. Instead of activating one cell, you click and drag over the cells so that a range of cells is now marked. Pasting is the same, except that the upper left cell of the group is the one whose location you specify as the target for the cut or copy.

Moving material between worksheets is accomplished the same way. You mark, indicate the cut or copy, move to the new sheet and the location on that sheet, and then paste.

- **Dragging and dropping**. If you want to cut or copy a range of cells on the same sheet, a shortcut to use is called **drag and drop**. Mark the cells, move the cursor to an edge, where it will become an outline arrow, hold down the mouse and drag to the new location. That is for cutting. To copy, hold down the **CTRL** key at the same time. If you attempt to drag and drop material onto a spot that currently has data in it you will get a message that asks if you want to do this, because it will remove the old material. If you have mistakenly done this, you can undo the change by clicking on the Undo icon, , that you will find in the Quick Access Toolbar. For more information, look up *drag and drop* in the help menu.

P.6 Formatting Numbers

Start with a clean spreadsheet. You can clear everything from a sheet by marking the whole sheet (Press CTRL+A) and then pressing the Delete key. Or, you can simply move to another sheet by clicking on a tab at the bottom of the screen. Enter these numbers into A1 through A6 and then copy them into column B, C, D, and E.

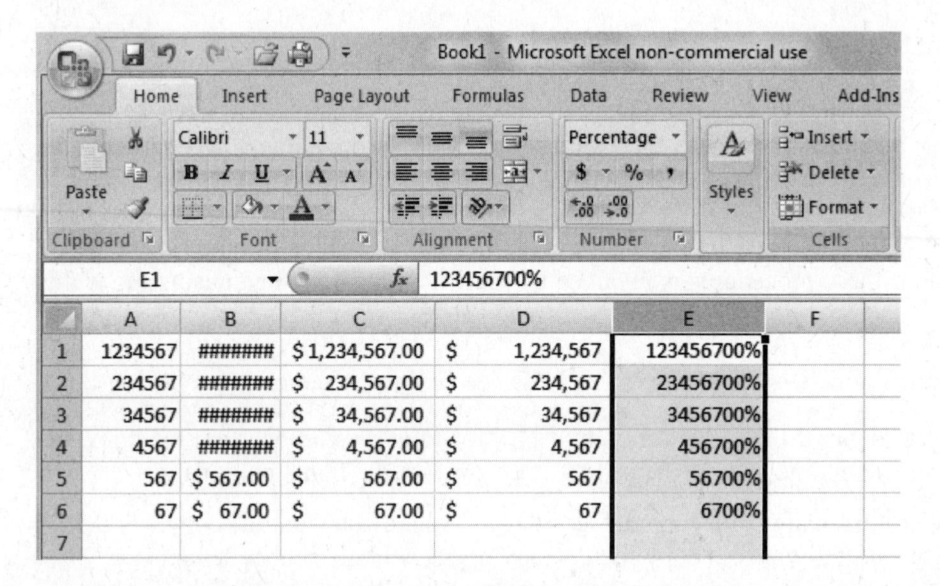

- **Currency** Click on the B at the top of the second column to mark it. In the Number group of the Home tab, you will find a $ icon. Click on it. The initial column width is set at 8.43. With this width, column B looks like the one shown above; all but the last two figures are replaced by ###. When you see this, it indicates that the numbers are too long to fit in that width. We can readjust the column width by double-clicking between the B and C at the top of the column. To show how this will change the appearance, we will activate column C, click on the $ icon and then double-click on the vertical separation between C and D to widen it an appropriate amount. The figure shows what results.

- **Currency variations.** If you wish to change to the format, which has no decimal, click on +.0/.00 or .00/→.0 icons. These options allow us to either increase or decrease the number of decimals to be displayed. In column D we clicked the .00/→.0 icon twice to get the column of values that eliminated the cents from being displayed.

- **Percent and comma** This option is located to the right of the **Currency** option. Click on column E and then click the % icon and note the changes. A number like 567 becomes 56700%.

P.6.1 Aligning Information

Start with a clearly unaligned set of data, such as the following:

A7 | Name Box | *fx* 0.1234567

	A	B	C	D	E
1	1234567				
2	123456.7				
3	12345.67				
4	1234.567				
5	123.4567				
6	12.34567				
7	0.123457				
8	1.23E-05				

Notice what happens when you enter cell A7. Entering the value .1234567 displays the value 0.123457. The six is missing. The program has been set to display digits to six places to the right of the decimal, if there is space. It rounds off any values that exceed that size. In cell A9, we entered the value .0000123456bu Excel displays the value as 1.23E-05, the exponential form of the number. We will not be concerned with these in this book, but they are commonly used in scientific measurements.

A simple way to align the eight entries above is to mark them and then click on the comma icon ,. Commas will be placed appropriately and values shown to two decimals. The last entry is displayed as 0.00. As above, the Formula Bar will display what we entered, .1234567, and that is the figure that will be used in computations. You can increase or decrease the number of decimals displayed, as described previously.

	A
1	1,234,567.00
2	123,456.70
3	12,345.67
4	1,234.57
5	123.46
6	12.35
7	0.12
8	0.00

P.6.2 Formatting a Range

Format a range of cells by marking it and applying the formatting style as was just demonstrated using individual cells. All cells within the range will be formatted in the selected style.

P.6.3 Inserting or Deleting Rows and Columns

Inserting or deleting rows and columns is relatively easy. If you have entered data in a row that includes columns A and B, and want to place a new column between A and B, do this: Click on the column B to mark that column. Right click on the mouse. A host of options becomes available to you. Select Insert and a new column will be created between the old columns A and B. Notice that the data contained in old column B will now be located in column C.

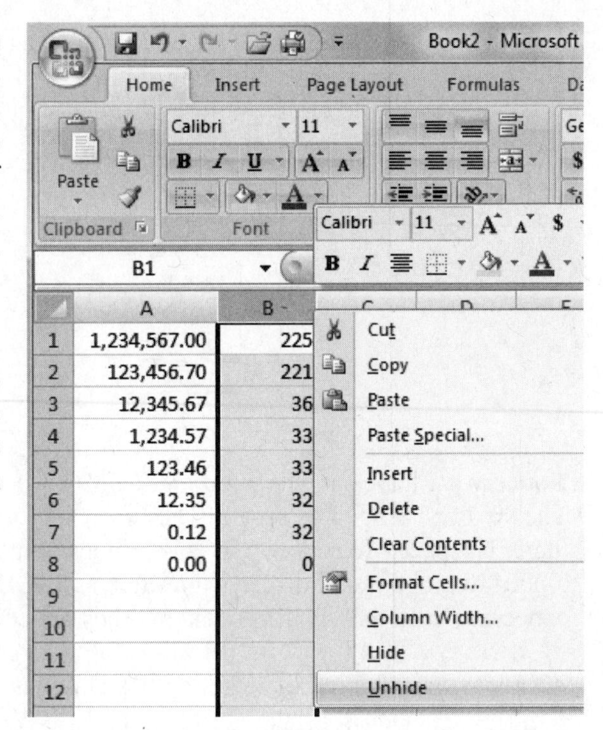

If you are starting over and want to insert two columns between A and B, simply mark columns B and C at the same time and follow the procedure above. To delete these two empty columns, mark them, right click on the mouse and click Delete.

The procedure for inserting or deleting rows is exactly the same, except that you use the numbers for the rows instead of the column letters.

P.6.4 Filling Adjacent Cells

Filling can mean one of two things. Filling can mean that you take the contents of a given cell or range of cells and make copies of that material in adjacent cells. Filling, broadly defined, can also mean that you continue a series or sequence of content into adjacent cells. For example, you might want to list the day of the week, and begin by typing Sunday into cell A1 and Monday into cell A2. You then can continue with Tuesday through Sunday in cells A3 through A8 with a simple move. We will discuss this in the section on **Series**.

Filling the same content into adjacent cells:

To learn how to fill the same content into adjacent cells, follow the steps below:

- On a clean spreadsheet, enter Sunday in cell A1 and Monday in cell A2.
- Click in the center of cell A1 and drag to cell A7, activating that range.
- In the Editing group of the Home Tab, click on the Fill icon, and click on *Down*.
- Now you will have Sunday copied in all the cells from A1 through A7. Copying takes the contents of the first cell in a series of marked cells and duplicates throughout the marked cells. The content of the second cell, Monday, is removed.

Filling a Series into Adjacent Cells

To learn how to fill a series into adjacent cells, follow the steps presented below:

- As above, enter Sunday in cell A1 and Monday in cell A2.
- Make sure Monday is entered by tapping Enter or clicking on the green check mark.
- Click in the center of cell A1 and drag through A2 so both cells are marked.
- Place the mouse pointer on the black dot (the **fill handle**) at the lower right of the two cells.
- The mouse pointer changes to a solid black plus sign. Drag it to cell A7.

The **series** of days from Sunday through Saturday is listed.

An easier method of accomplishing the same thing is presented below:

- Use the above layout of Sunday in cell A1 and Monday in cell A2
- Click on A1 to activate it.
- Grab the fill handle of cell A1 and drag to A7 to complete the series.

Days of the week are special, since we have a given order determined by convention. If we choose numbers that have no agreed upon sequence, we will not get a series generated by simply choosing the first number on the list.

We can also generate a series of numbers, but the series must have an identifiable pattern.

- On a clean spreadsheet, enter 10 in A1 and 15 in A2.
- Mark A1 and A2 by dragging over them.
- Using the fill handle, drag to A10
- You should have the numbers 10, 15, 20, 25, … to 55 listed.

P.6.5 Series

The above activities show how easily we can copy the contents of a given cell into adjacent cells or generate a series of numbers from two examples. Now we will use the **series** command to generate a series of numbers.

Follow the steps presented below.

Enter these numbers in cells A1:A3: 5, 7, 9.
Activate the range A1: A10.

Click on the Fill icon. Then Choose the Series option. It will produce the dialog box shown below that provides information about the series for you to verify before the program continues the series.

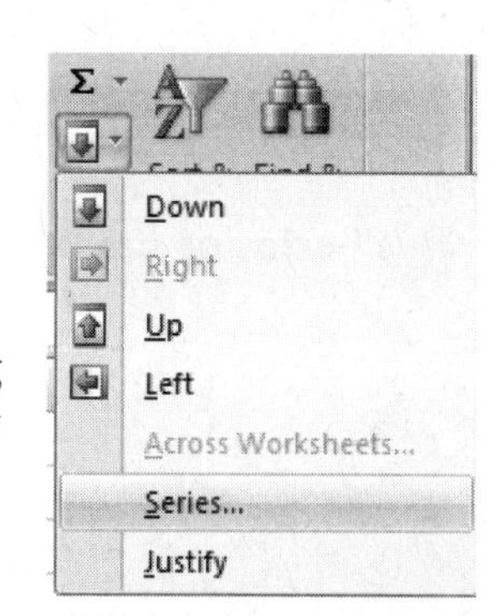

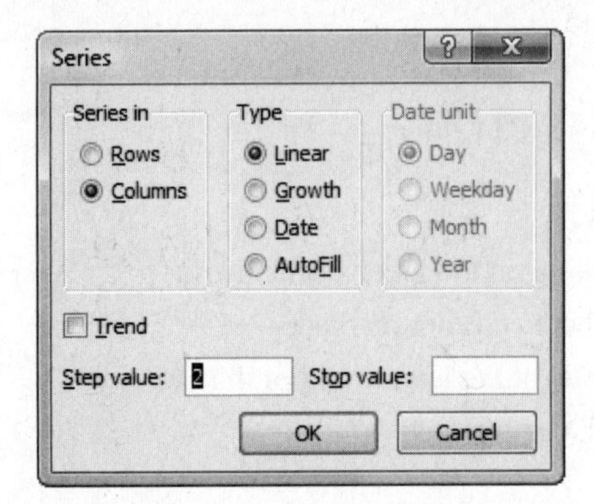

The dialog box displays the program's interpretation of the data. The series is presented in **Columns**. It is a **Linear** relationship. No **Date Units** are provided. The **Step Value** and **Stop Value** are available for verification and determination. If you wanted to specify a value to stop at, it would be entered in that box.

P.6.6 Sorting

Sorting becomes a useful tool when we have a set of data to analyze and want an initial look to see how it is distributed. There are several methods of sorting data that we will describe. Enter the data as shown below into a fresh spreadsheet and click and drag the mouse to select the data.

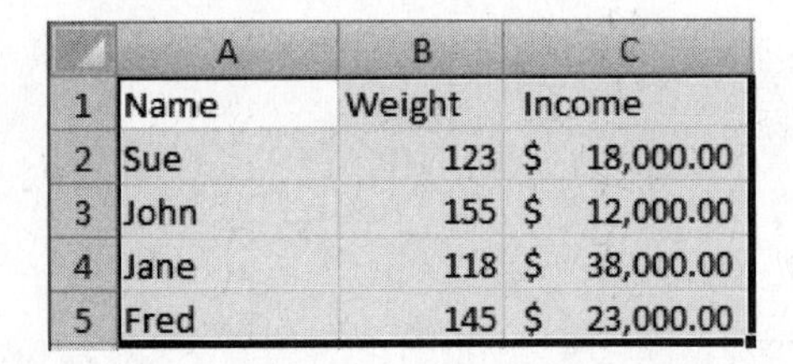

	A	B	C
1	Name	Weight	Income
2	Sue	123	$ 18,000.00
3	John	155	$ 12,000.00
4	Jane	118	$ 38,000.00
5	Fred	145	$ 23,000.00

For a quick sort of the data, click on either the AZ↓ or ZA↓ icons located in Sort & Filter group of the Data tab. These icons will sort the data by the variable in the first column selected either in ascending or descending order. By clicking on the AZ↓ icon, the data will be sorted by Name in ascending order. The results of the sort are shown below.

	A	B	C
1	Name	Weight	Income
2	Fred	145	$ 23,000.00
3	Jane	118	$ 38,000.00
4	John	155	$ 12,000.00
5	Sue	123	$ 18,000.00

By clicking on the ZA↓ icon, the data will be sorted in descending order.

More advanced sorting options are available when the sorting icon, 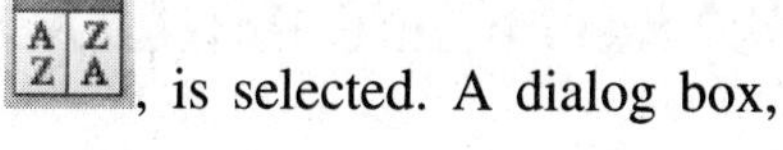, is selected. A dialog box, shown below, requires that we specify which column to use in sorting:

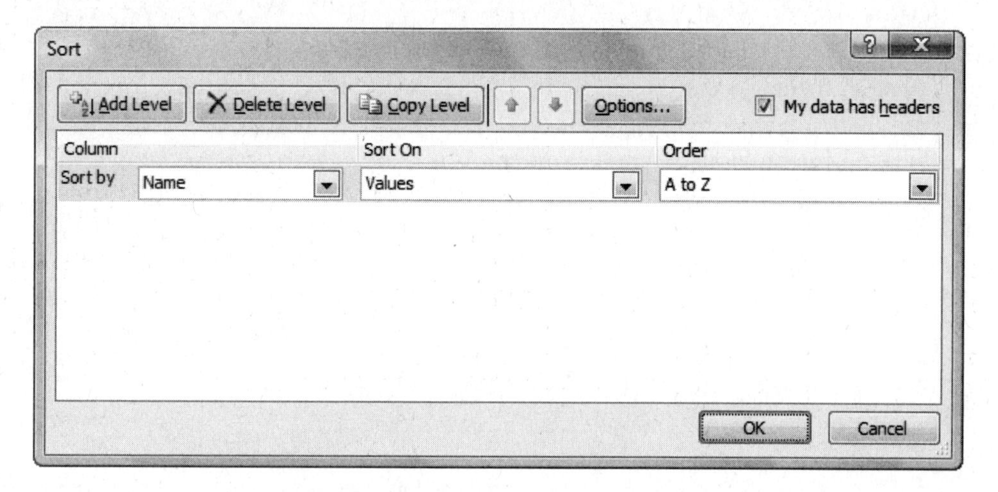

Using this option, we have more choices as to which columns we wish to use as the sorting variable. The upper left cell, activated when we marked the range, is listed as the default variable to use. To select a different variable, we pull down the Column tab and select a different variable. Suppose we select to sort based on the Weight variable. The resulting output is shown below.

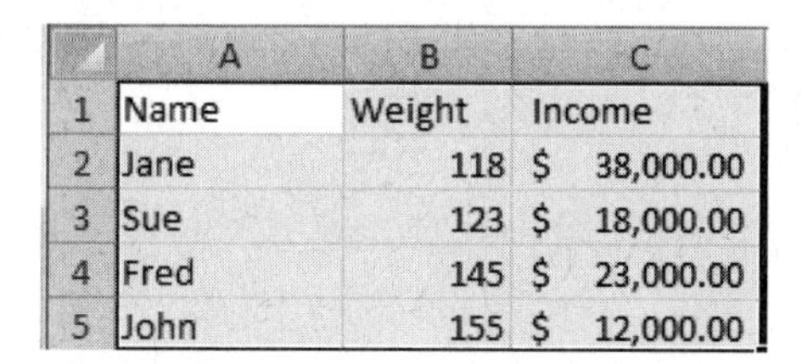

	A	B	C
1	Name	Weight	Income
2	Jane	118	$ 38,000.00
3	Sue	123	$ 18,000.00
4	Fred	145	$ 23,000.00
5	John	155	$ 12,000.00

Notice the data is sorted by ascending Weight. Our last option would be to sort based on the Income values in the data set.

By clicking the Add Level button in the sort menu, we have the option of sorting based on additional variables. If, for example, we had a list of 100 student members of an organization that included active and former membership status and address (street, city, zip code), then we would have a number of zip codes that are identical. In the case we might want to sort by zip code, and then by membership status if we were generating mailing labels (sorted by zip code for a cheaper rate) and different messages for different types of members. The one identifier that should not have any duplicates is social security account number, which is why it is often used as an identifier.

P.7 Saving and Retrieving Information

P.7.1 Naming Workbooks

Earlier versions of Excel used one sheet as the basic unit that was stored as one file. Now, many different worksheets (determined by system memory) can be saved in a single workbook. If you have opened a new workbook and entered data that you want to save, the logical next step is to click on save icon, , to save your work. When you have a new, previously unnamed sheet, you're then prompted to give the file a name. The default name of Book 1 appears in the File Name window of the Save As dialog box. Unless you indicate otherwise, the file will be saved in the last location used, be it the hard drive, network server, or diskette. A common default is to save in a folder called *My Documents*. In the figure below, I have chosen to name the workbook Primer Data Set and save it in a folder that I have created called Excel Manual 11th Edition

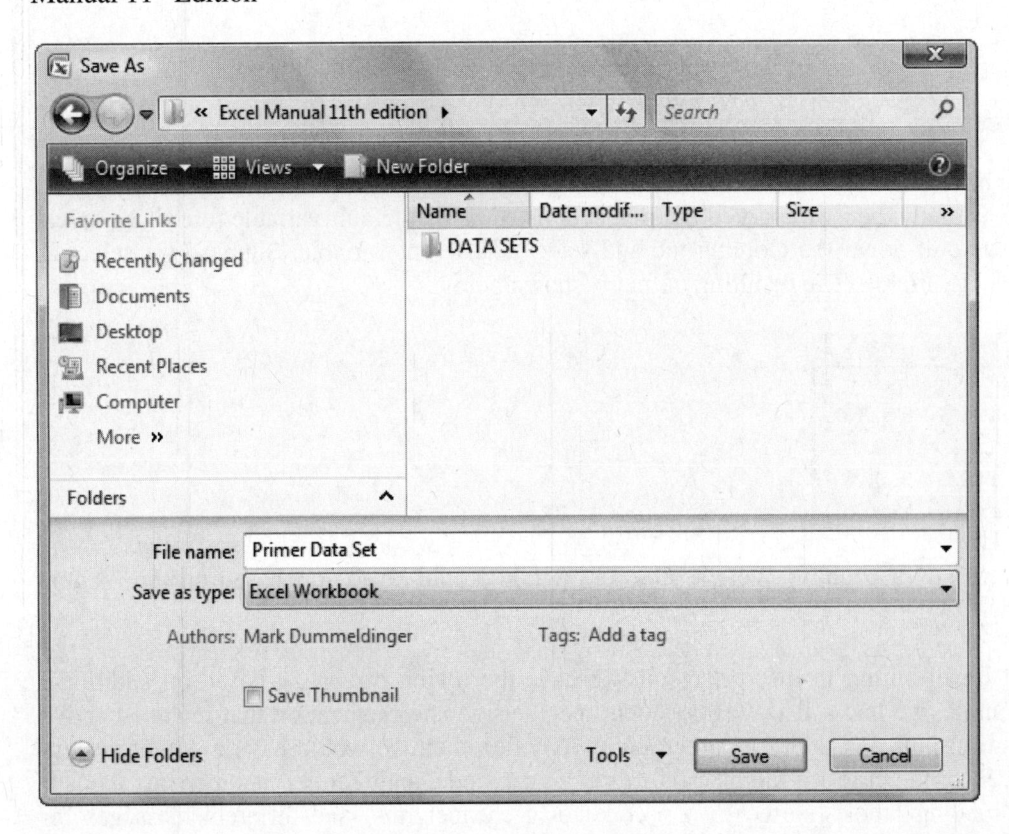

In most current versions of Excel, files can have names longer than eight letters. In earlier versions, the **DOS** naming convention applies. In that case, a file cannot have a name longer than eight letters, followed by a period and an extension, usually .xls. There are restrictions on the symbols that you can use in a file name, which will be apparent when you get an error message for using a/or some other forbidden symbols. Newer Windows versions and earlier Mac versions allow names to have up to 256 characters, so they can be more descriptive of the contents. Use the *?* and *Help* to learn the names and functions of the various boxes in the *Save As* dialog box.

P.8 Printing

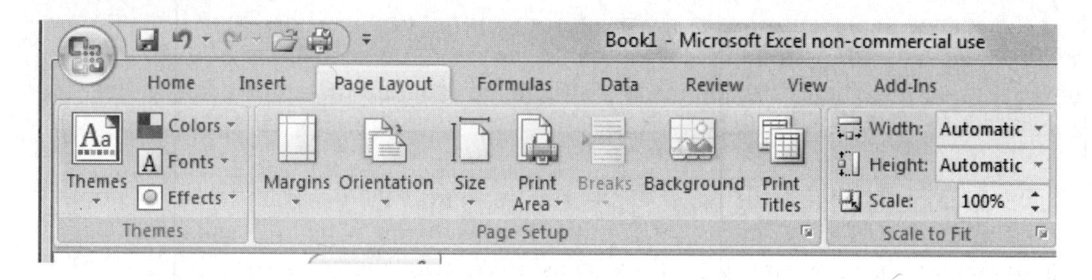

P.8.1 Page Setup

The **Page Layout** tab includes the page options that are most useful for getting the correct look of the worksheet. Rather than look at all of these options individually, we will summarize the more important options by clicking on the expand arrow, , located in the lower right-hand corner of the Page Setup group. We will go through each of these tabs one at a time.

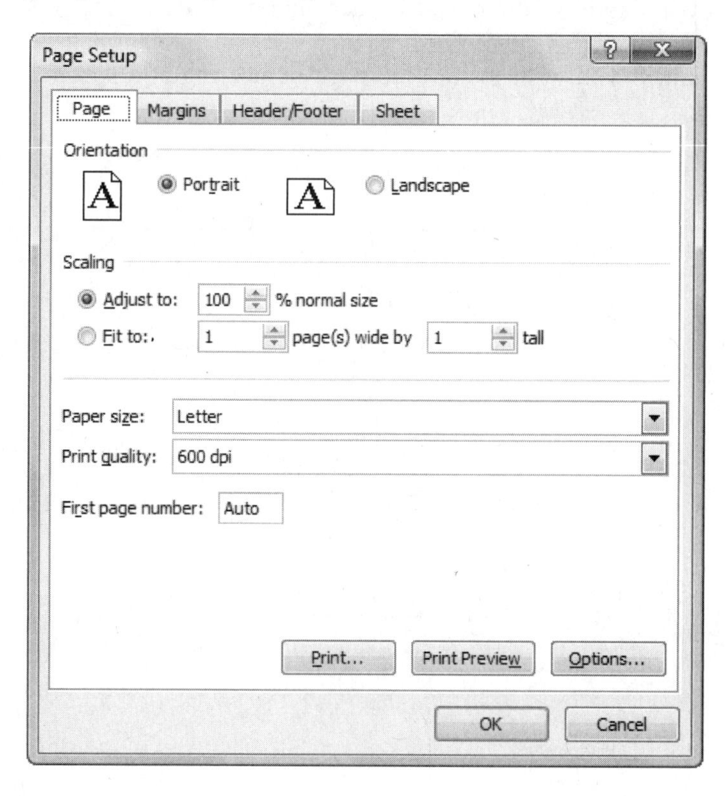

P.8.2 Page

Page allows the user the change the **orientation** of the page from portrait (the way this book is printed) to landscape (sideways). With the **scaling** commands you can adjust the size of the image to be 10% of normal size or enlarge it to 400%. The **Fit to** command instructs the program to automatically adjust the size to fit any of a number of pages as specified.

> Beware: If you have three or four pages of material to print and direct the program to fit it to one page, it is sometimes difficult to read. You can specify that it will stay one page high and allow it to continue to the right on other sheets as far as necessary by using the two adjustments available.

Paper Size and **Print Quality** provide options shown via the drop-down arrow, although most often letter size paper is used, and print quality adjustments are not available. On some printers, you can move to a higher quality print by changing the dpi to a value of 1200 or more. This does increase printing time and use of ink.

Clicking on **Options** provides more graphic views of some of these choices, depending on your printer.

First Page Number is set to **Auto** as the default. If you wish to start the page numbering at a specific value, type in that value after highlighting **Auto**.

P.8.3 Margins

Margins are set using the options shown by clicking the Margins tab. The preview window shows how these will look as you change them. Note that you can also center the output on the page horizontally, vertically, or both. After you have changed any of the settings you can click on **Print Preview…** to check its appearance. You can also change the margins in Print Preview by dragging the margin boundary handles.

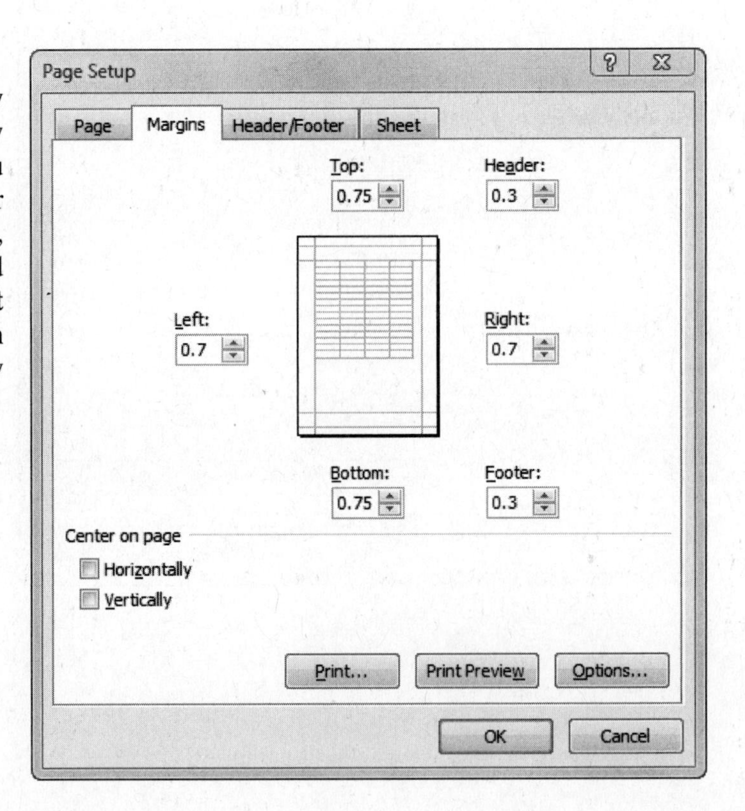

P.8.4 Header/Footer

Headers contain the information printed across the top of all (or all but the first) pages. In this text the headers on one page indicate the chapter title and page number; on the other we have the section and page number. Sometimes this information is printed at the bottom of a page, and called a **Footer**. Our description of the procedures for developing headers also applies to footers.

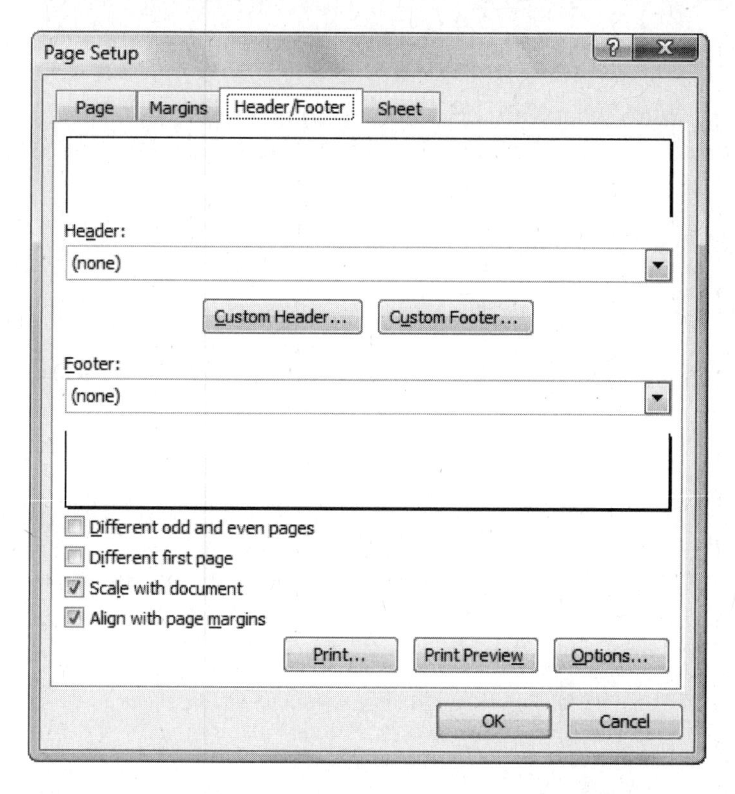

Unless you alter the settings, the **header** and **footer** shown above are the default. Click on the arrow to the right of the default header and you will see some previously used on the computer. You may wish to choose one of them and edit it to your liking. Or, if you wish, you can click on **Custom Header…** and obtain the following dialog box:

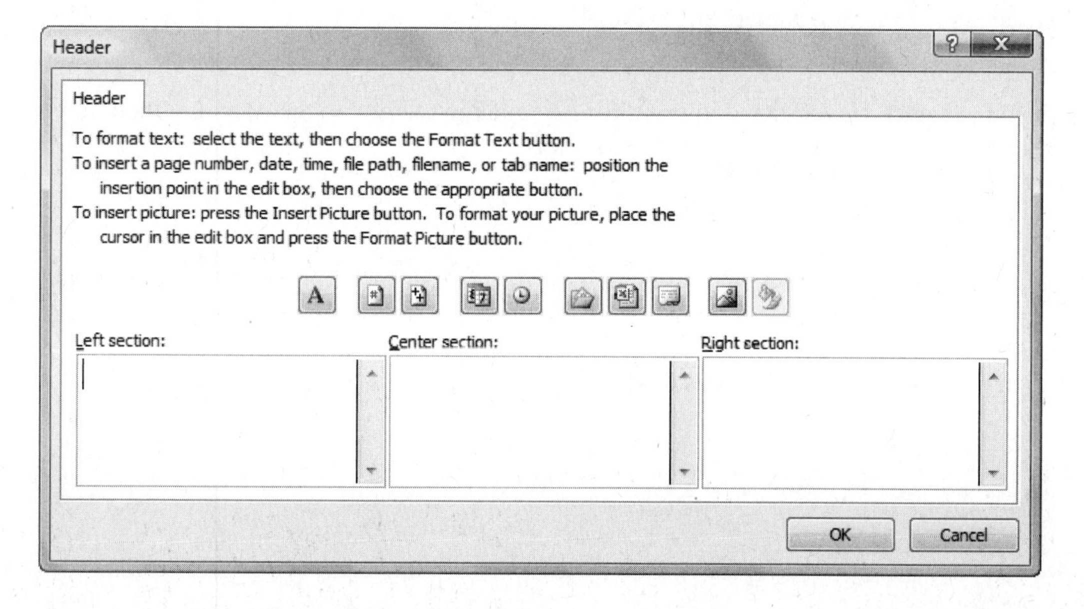

In this case, you have the option to place your header in the left, center, or right of the page. Scrolling the mouse over the various icons will give you a brief description of what each will do for you. The A icon, when clicked, allows you to write in text and change the font, size, and style of the text ; others are for page numbering, date and time, filename and sheet name.

P.8.5 Sheet

This page allows you to choose the area to print, select whether to print titles of columns or rows on every page printed, print gridlines, select the quality of print, and choose whether to print row and column headings.

Print Area allows you to select a given range to print. Simply click in the box to the right and then move to the spreadsheet where you drag to indicate the range to select.

P.8.6 Enhancing Output

With a bit of experience you can improve the appearance of your output by using the options described above. Using the preview option allows you to see exactly how things will look before you print them. Sometimes gridlines improve the display. An enlarged title adds clarity. Size of the sheet can be adjusted for your purposes. All of these can be explored using the options above.

P.8.7 Inserting and Removing Page Breaks

When a worksheet spans more than one page, Excel determines where to insert a page break. You might find it more convenient to make the break at a different location. Use the Breaks icon located in the Page Setup group of the Page Layout tab to set page breaks in the worksheet. Print Preview, which is explained next, is always a good way to determine what looks best prior to printing your worksheet.

P.8.8 Preview and Print

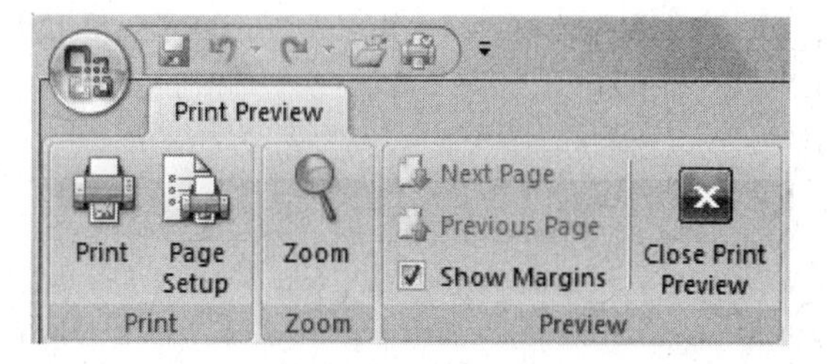

Clever computer users preview their material before they print it. You can go to the Print Preview command by choosing the Print option from the options listed under the Microsoft Office Button. To change options, after you view your sample, you can use the menu at the top. Try out such things as Zoom, which will increase the size of the page so you can view details. Click it to turn it on and click it to return to the full page view. Setup will display the four pages described above, in that section. By checking the Show Margins box, Excel will display lines on your preview that shows where the margins will be located. You can grab the handles and change them on the sheet. Close Print Preview returns you to the regular view of the page. Print moves you to the dialog box with choices regarding the printer to use, number of copies, which pages to print, etc.

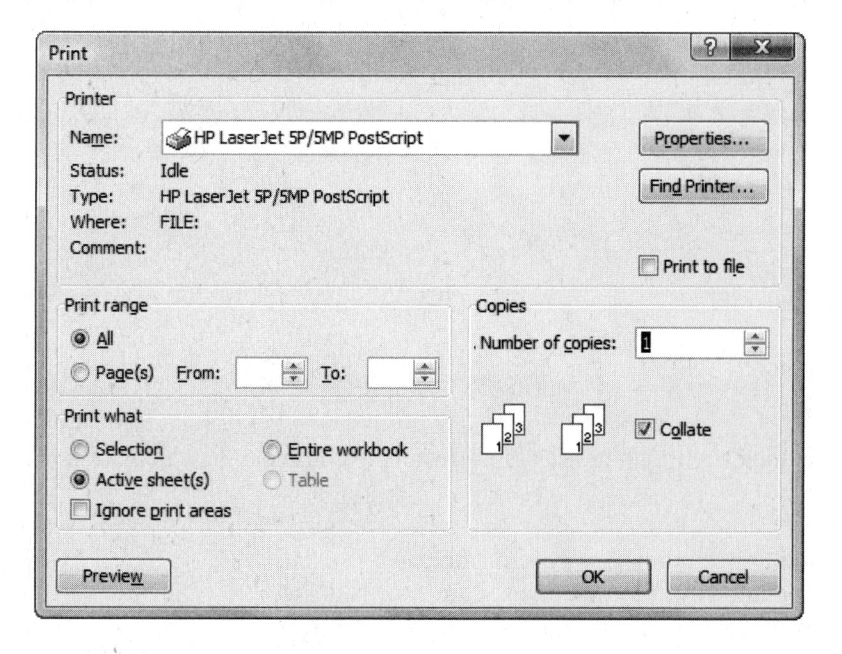

P.9 Using Formulas and Functions

P.9.1 Operators

Algebraic formulas utilize the four common operations of addition (+), subtraction (–), multiplication (×), and division (÷). When creating formulas, we will use these operations along with three others. Called **operators**, they are symbolized as:

Addition + Multiplication * Subtraction – Division /

Negation refers to using the minus sign to indicate a negative number, as in –3

Exponentiation ^ Percent %

Multiplication uses an asterisk instead of an ×. Division uses a diagonal (/) instead of the division symbol. If we wish to raise a value to a power, say X^2 we place the carat (^) between the X and the 2. To raise a number X to the second power, we would write X^2. Finally, we can also place a percent (%) sign behind a value, as in 20%. For example, the formula =15 ^ 2 * 15% raises 15 to the second power and multiplies the result by 0.15 (the decimal form of 15%) to produce the result of 33.75.

We also group sets of operations together with parentheses, which determine the order of execution of commands by the computer. Any operations enclosed in parentheses are executed first, moving from the innermost parentheses to the outermost.

P.9.2 Order of Operations

The order of operations in Excel is:

1. Negation, as in –15
2. Percent
3. Exponentiation
4. Multiplication and division
5. Addition and subtraction

Excel first calculates expressions in parentheses and then uses those results to complete the calculation of the formula. For example

= 2+4*3 produces 14 because multiplication occurs before addition
=(2+4)*3 produces 18 because operations within parentheses are executed first
=2+(4*3) produces 14 because operations within parentheses are executed first

In a more complex situation,

=2+(4*(3+5)^2)/2 produces 130.

First the terms in the innermost parentheses are executed, producing 8. Then this is raised to the second power, producing 64, since this is within the second set of parentheses and exponentiation has precedence over the other operations. 64 is multiplied by 4, yielding 256. This is divided by 2, producing 128. Adding

2 to this amount yields the final answer, 130. Try this on your computer, by entering this formula into cell A1 and pressing Enter.

P.9.3 Writing Equations

In the equations above, we have always used numeric constants, such as 2, 3, 4, or 5. In practice, the formula that you write will probably refer to a cell that may contain any value. This is equivalent to the X and Y that acted as unknowns in algebra. For example, to convert degrees Fahrenheit to Celsius, we use the formula:

C = 5/9(F-32)

Let's set up a table to provide this conversion. Type the label in cell A1: *Degrees Fahrenheit.* In cell B1 type: *Degrees Celsius.* Enter the two values of –60 and –59 in cells A2 and A3. That's a good place to start if you live in Wisconsin in January.

	A	B
1	Degrees Farenheit	Degrees Celsius
2		
3	-60	
4	-59	
5		

Let Excel continue the series by clicking on the value of –60 and dragging down to cell A80 to mark those cells. Then go to the Fill icon, as described earlier, to continue the series by adding one degree to the temperature with each successive cell. Notice that the program has correctly inferred that the step size is +1. Each time we move down a row the value in column A is reduced by one unit. We wish to continue the series with that change. Click OK and we have data extending from –60 to 17 degrees. In cell B3 enter this formula:

=5/9*(A3-32) and press Enter.

When the temperature is –60 degrees Fahrenheit, it is –51.11 degrees Celsius, as cell B3 indicates. Activate cell B3. Now, just click on the small square in the lower right of cell B3 (the fill handle), and drag to cell B80. The screen pointer becomes a solid black plus sign when you use the fill handle. You have copied the formula into all of those cells, with the address of each cell that represents degrees Fahrenheit changing for each row. You should have something like the following:

To make our output more attractive, click on column B to activate that column, go to Format – Cells – Number and accept the default of two places to the right of the decimal. Now you should have the first few rows of data looking like this:

	A	B
1	Degrees Farenheit	Degrees Celsius
2		
3	-60	-51.11111111
4	-59	-50.55555556
5	-58	-50
6	-57	-49.44444444
7	-56	-48.88888889
8	-55	-48.33333333
9	-54	-47.77777778
10	-53	-47.22222222
11	-52	-46.66666667
12	-51	-46.11111111
13	-50	-45.55555556
14	-49	-45

P.10 Entering Formulas

Now, let's take a situation that is very commonly used by students. Compute a Grade Point Average. Enter the following information in a new blank spreadsheet.

	A	B	C	D	E
1	Course	Credits:	Grade:	Value:	
2					
3	English	5	A	4	
4	Math	4	B	3	
5	Psychology	3	A	4	
6	Economics	3	C	2	
7		15			
8					
9	Grade:	Value			
10	F	0			
11	D	1			
12	C	2			
13	B	3			
14	A	4			
15		5			

We want to compute the Grade Point Average. We have to multiply the credits for a course by the value of the grade given for that course. In Cell E3 enter this formula: =B3*D3. Then use the fill handle in the cell to copy the formula from E3 to E6, so that you have the product of credits multiplied by the value of the grade for each course. The recommended procedure for setting up formulas is to click on the cell whose address you want entered, NOT type in the address of that cell. In this case you would type the = sign in E3 and then click on B3, which now appears in the formula. Then enter an asterisk (from either SHIFT-8 or the * key on the keypad) followed by clicking on D3. Then either click on Enter or the green arrow in the formula bar. You should have:

	A	B	C	D	E
1	Course	Credits:	Grade:	Value:	Grade Points
2					
3	English	5	A	4	20
4	Math	4	B	3	12
5	Psychology	3	A	4	12
6	Economics	3	C	2	6
7		15			
8					
9	Grade:	Value			
10	F	0			
11	D	1			
12	C	2			
13	B	3			
14	A	4			
15		5			

Now all we need to do is to sum the Grade Points, sum the Credits, and then divide Grade Points by Credits to get GPA. Use the AutoSum button, Σ a Greek Sigma on the Standard Toolbar) to obtain the sum for each. Just activate cell B7 and click on that button. A series of dotted lines surrounds the four digits above, indicating that the program estimates that this is what you want to sum. Click Enter and the value of 15 will be displayed. Do the same for the Grade Points to obtain 50. To obtain the GPA you can enter those two values in an equation directly or enter the cell references in the equation to obtain the answer. If you do the latter, your equation would look like: =E7/B7. Put your answer in cell E9 as shown below:

	A	B	C	D	E
1	Course	Credits:	Grade:	Value:	Grade Points
2					
3	English	5	A	4	20
4	Math	4	B	3	12
5	Psychology	3	A	4	12
6	Economics	3	C	2	6
7		15			50
8					
9	Grade:	Value		GPA=	3.333333333
10	F	0			
11	D	1			
12	C	2			
13	B	3			
14	A	4			
15		5			

Note that the value of 3.3333… doesn't appear until you press Enter of click on the green arrow. If you click on cell E9 now, the formula you used to obtain that value will be displayed in the Formula Bar, not the value shown here. For emphasis, you can mark cells D9 and E9 and then click on the symbol to make them **BOLD**.

P.10.1 Relative References

When we entered the formula for determining Grade Points in cell E3, we entered it using the relative reference mode. We told the program to go to cell B3, obtain that value, go the cell D3, obtain that value, multiply them and place the answer in E3. Actually we instructed Excel to move the three cell locations to the left (B3) of the active cell, obtain that number, move one cell to the left (D3) of the active cell, obtain that number, multiply them and place the answer in the active cell. To see how this works, try this: Activate cell E3. Note the formula in the Formula Bar. Now activate cell F3. The formula references in the formula bar have been adjusted to reflect what was said above; the program is not going to absolute locations of cells, but is moving relative to the active cell, using the same movements in the spreadsheet that were specified in the formula. Now let's make our computation of the GPA a bit more sophisticated by using that table that gives the value of each grade. We will use a function called **table lookup**.

P.10.2 Absolute References

When we have a table that converts a letter grade to a corresponding number (or vice versa), we can use a built-in function known as **VLOOKUP**. Mark cells D3 through D6 and press delete to clear the contents. We will have the program **VLOOKUP** compare the grade letter printed in cells C3 through C6 to the tabled values in cells A10:B14 to fill in the number value corresponding to a letter grade. Click on the Formulas tab and click on the Function icon, shown here. Move to the **Lookup & Reference** part of the Paste Function Window as shown below. Click on **VLOOKUP** as indicated. Then click OK.

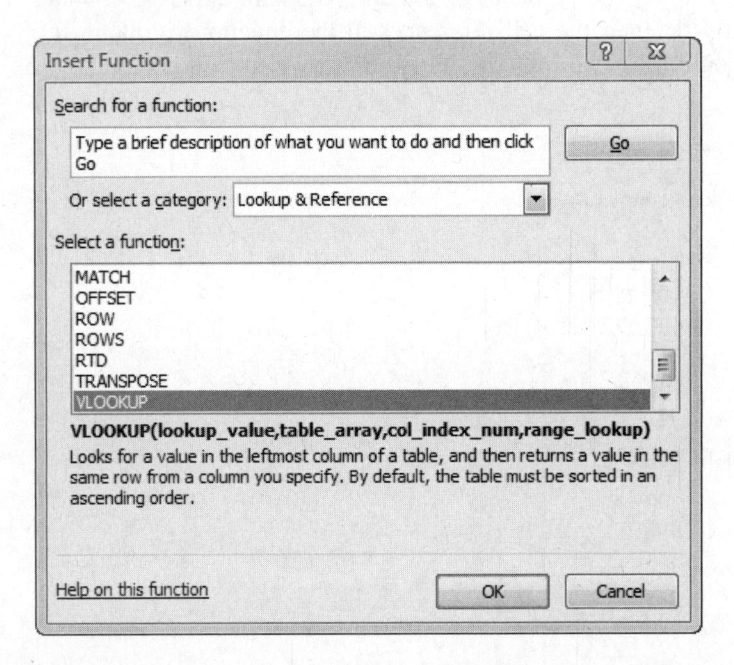

The following window appears after you execute the above and enter the values into the formula as shown in the formula bar. After we enter the =VLOOKUP in cell D3 we enter the location of the grade to translate (cell C3), then the absolute address of the table ($A$10:$B$14), the letter 2 to indicate that we want the number from the second column in the table, followed by FALSE which indicates that we want an exact match. After accepting the formula by clicking the green arrow, we use the fill handle to copy the formula into cells D4: D6. The cells $A$10:$B$14 are absolute references because we do not want them to change as we refer to them in the operations on the second, third, and fourth grades. Whenever we want an address to be absolute, we preface the usual column letter and row number with a dollar sign.

The figure below shows the formula entered into cell D3 and the resulting value of 4. Note how the C3 value is a relative value while references to the table are absolute.

D3 | f_x =VLOOKUP(C3,A10:B14,2,FALSE)

	A	B	C	D	E	F
1	Course	Credits:	Grade:	Value:	Grade Points	
2						
3	English	5	A	4	20	
4	Math	4	B	3	12	
5	Psychology	3	A	4	12	
6	Economics	3	C	2	6	
7		15			50	
8						
9	Grade:	Value		GPA=	3.333333333	
10	F	0				
11	D	1				
12	C	2				
13	B	3				
14	A	4				
15		5				

Chapter 1
Statistics, Data, and Statistical Thinking

1.1 Introduction

Chapter 1 introduction the topics that will be expanded on throughout the text. No data analysis is necessary in Chapter 1 and Microsoft Excel® cannot be used here in the text.

Chapter 2
Methods for Describing Sets of Data

2.1 Introduction

Chapter 1 served to introduce many of the basic statistical concepts employed in all types of data analysis problems. Two main areas of statistics emerge from Chapter 1 - descriptive and inferential statistics. Chapter 2 focuses on the descriptive area and looks at both graphical and numerical techniques that allow statisticians to summarize data that has been collected. Many of the techniques used to summarize data discussed in *Statistics for Business and Economics* can easily be performed with Excel. Our purpose is to explain these techniques and to illustrate them using the examples presented in the text as well as additional examples provided here. Listed below are the various techniques that Excel offers that can be used to generate the graphical and numerical topics presented in Chapter 2.

Excel offers a wide array of graphing options to the statistician. When working with qualitative data, Excel allows the statistician to create customized **pie charts** and **bar graphs**. For quantitative data, **bar graphs**, **box plots**, **histograms**, and **scatter plots** are easy to create. The scatter plot feature in Excel can readily be used to create the **time series plot** discussed in the text.

As with most database and statistical software programs, Excel provides a wide array of numerical description of data. The three measures of central tendency (**mean**, **median**, and **mode**) and the three measures of variability (**range**, **variance**, and **standard deviation**) are all available in the descriptive statistics menu of Excel. Measures of relative standing (**percentiles**, **quartiles**, and **z-scores**) are available in Excel but not as easy to access as the measures of central tendency and measures of spread.

The following examples from *Statistics for Business and Economics* are solved with Microsoft Excel® in this chapter:

Excel Companion Exercise	Page	Statistics for Economics and Business	Excel File Name
2.1	32	Example 2.2	Pricequotes
2.2	40	Table 2.1	Forbes40
2.3	42	Example 2.19	Medfactors
2.4	45	Example 2.2	Pricequotes
2.5	46	Example 2.4	R&D
2.6	49	Example 2.10	R&D
2.7	50	Example 2.14	R&D
2.8	52	Table 2.2	R&D

2.2 Graphical Techniques in Excel

2.2.1 Bar Graphs and Histograms

Bar graphs, pie charts, and scatter plots are all easy to generate using Excel. The graphs enable the user to summarize the data that they are viewing and make decisions quickly and easily. **The DDXL Add-In** offers an easy method of creating pie charts and bar graphs for qualitative data within the **One-Way Tables and Charts** menu. For quantitative bar graphs, we will utilize the **Histogram Data Analysis** procedure within the Excel program.

Exercise 2.1: As an example we turn to Example 2.2 from the *Statistics for Business and Economics* text:

A manufacturer of industrial wheels suspects that profitable orders are being lost because of the long time the firm takes to develop price quotes for potential customers. To investigate this possibility, 50 requests for price quotes were randomly selected from the set of all quotes made last year, and the processing time was determined for each quote. The processing times are displayed below in Table 2.1, and each quote was classified according to whether the order was "lost" or not (i.e., whether or not the customer placed an order after receiving a price quote).

Table 2.1

Request	Time	Lost?	Request	Time	Lost?
1	2.36	No	26	3.34	No
2	5.73	No	27	6.00	No
3	6.60	No	28	5.92	No
4	10.05	Yes	29	7.28	Yes
5	5.13	No	30	1.25	No
6	1.88	No	31	4.01	No
7	2.52	No	32	7.59	No
8	2.00	No	33	13.42	Yes
9	4.69	No	34	3.24	No
10	1.91	No	35	3.37	No
11	6.75	Yes	36	14.06	Yes
12	3.92	No	37	5.10	No
13	3.46	No	38	6.44	No
14	2.64	No	39	7.76	No
15	3.63	No	40	4.40	No
16	3.44	No	41	5.48	No
17	9.49	Yes	42	7.51	No
18	4.90	No	43	6.18	No
19	7.45	No	44	8.22	Yes
20	20.23	Yes	45	4.37	No
21	3.91	No	46	2.93	No
22	1.70	No	47	9.95	Yes
23	16.29	Yes	48	4.46	No
24	5.52	No	49	14.32	Yes
25	1.44	No	50	9.01	No

a. Use a statistical software package to create a frequency histogram for these data. Then shade the area under the histogram that corresponds to lost orders.

b. Use a statistical software package to create a stem-and-leaf display for these data. Then shade each leaf that corresponds to a lost order.

c. Compare and interpret the two graphical displays of these data.

Solution:

We answer part a. by utilizing the bar graph utility within the **DDXL** program. Before we begin, we must access the data set for this example. **Open** the Data File **Pricequotes** by following the directions found in the preface of this manual. If done correctly, the data should appear in a workbook similar to that shown below in Figure 2.1 Use the mouse to select the data shown in the workbook.

Figure 2.1

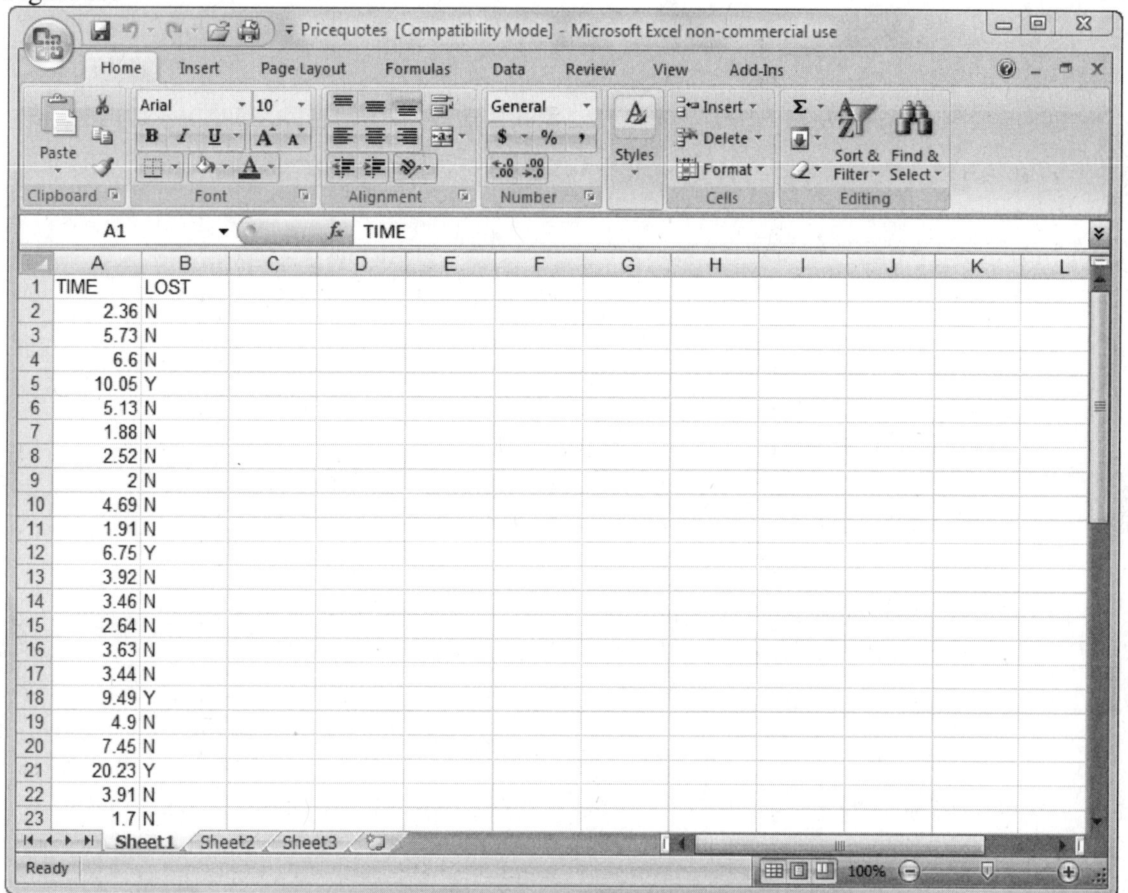

Our first goal in part a. is to create a bar graph for the lost variable in the data set. Click on the **DDXL Add-In** menu. Click on the **Charts and Plots** option to access the **Charts and Plots Dialog** menu (see Figure 2.2 below). Click on the ▼ in the Function Type box to access the different charts available to select. Highlight the **Bar Chart** option. Highlight the LOST variable found in the Names and Columns Box. Click on the ◀ to the left of the Categorical Variable box to create a bar chart for the Lost variable. Click **OK**. The bar chart is shown in Figure 2.3.

Figure 2.2

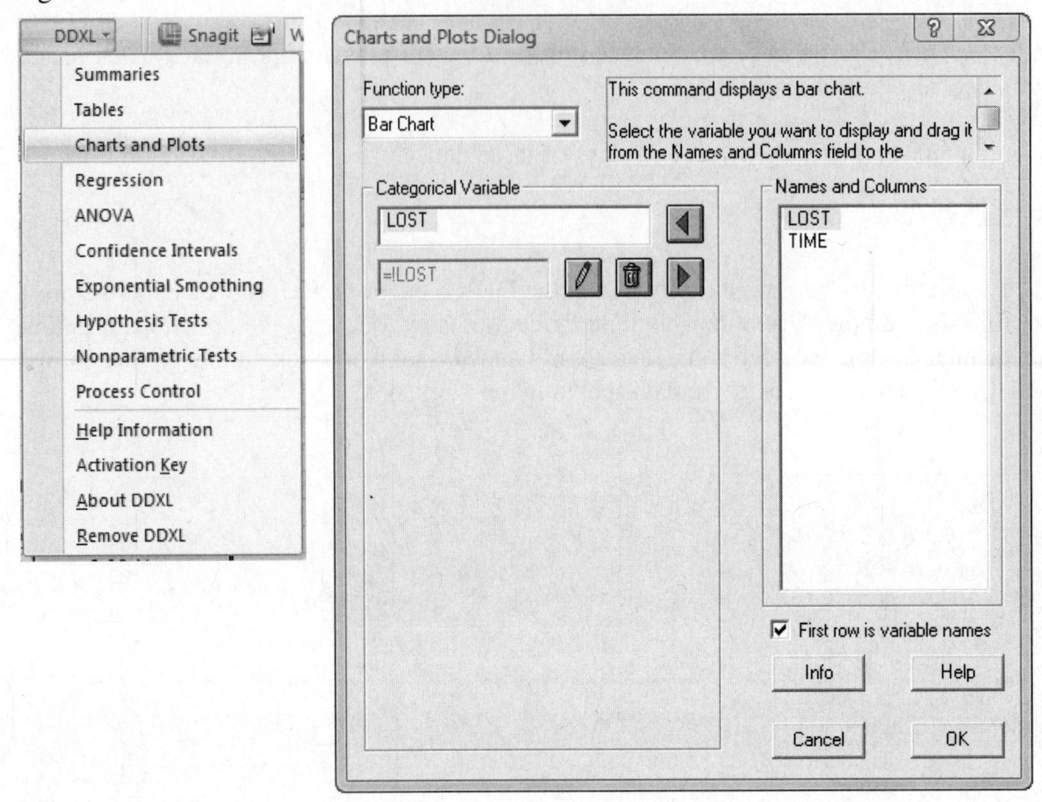

Figure 2.3

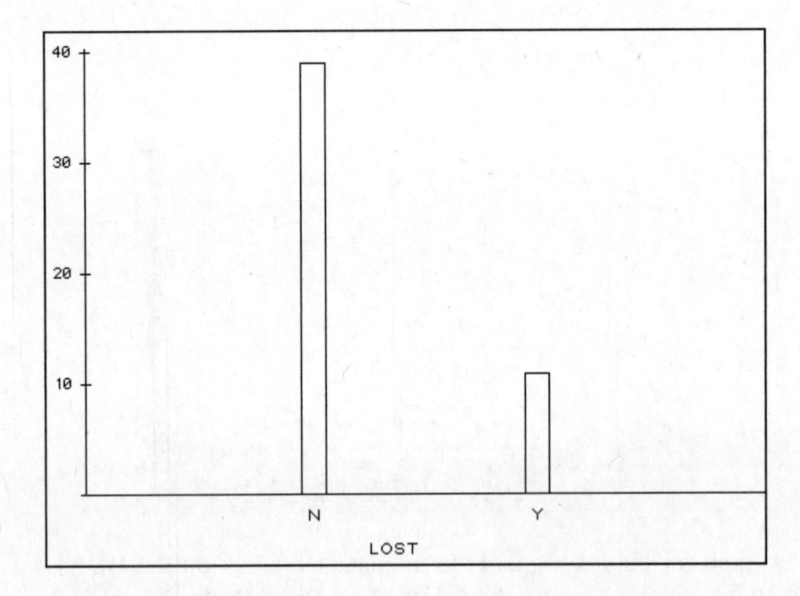

The next step is to generate a histogram for the quantitative variable, Time. To do this we begin by again clicking on the **DDXL Add-In** menu. Click on the **Charts and Plots** option to access the **Charts and Plots Dialog** menu (See Figure 2.4 below). Click on the ▾ in the Function Type box to access the different charts available to select. We highlight the **Histogram** option. Highlight the TIME variable found in the Names and Columns Box. Click on the ◀ to the left of the Quantitative Variable box to create a histogram for the Time variable. Click **OK**. The histogram is shown in Figure 2.5.

Figure 2.4

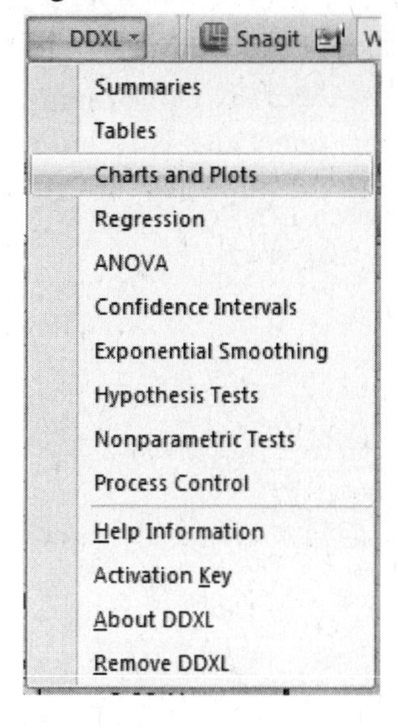

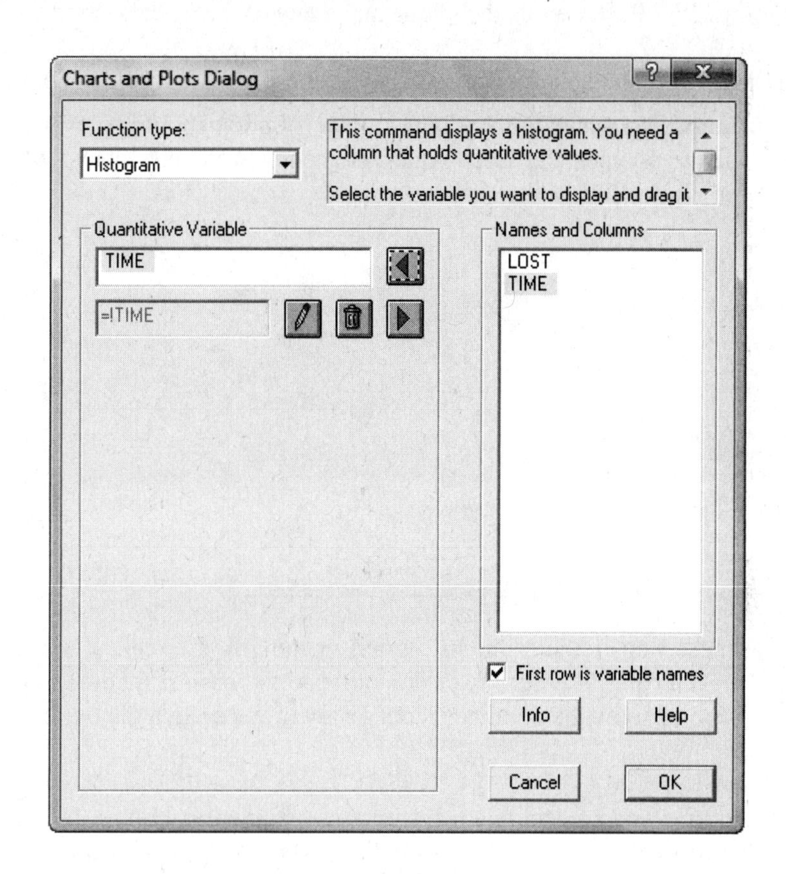

Figure 2.5

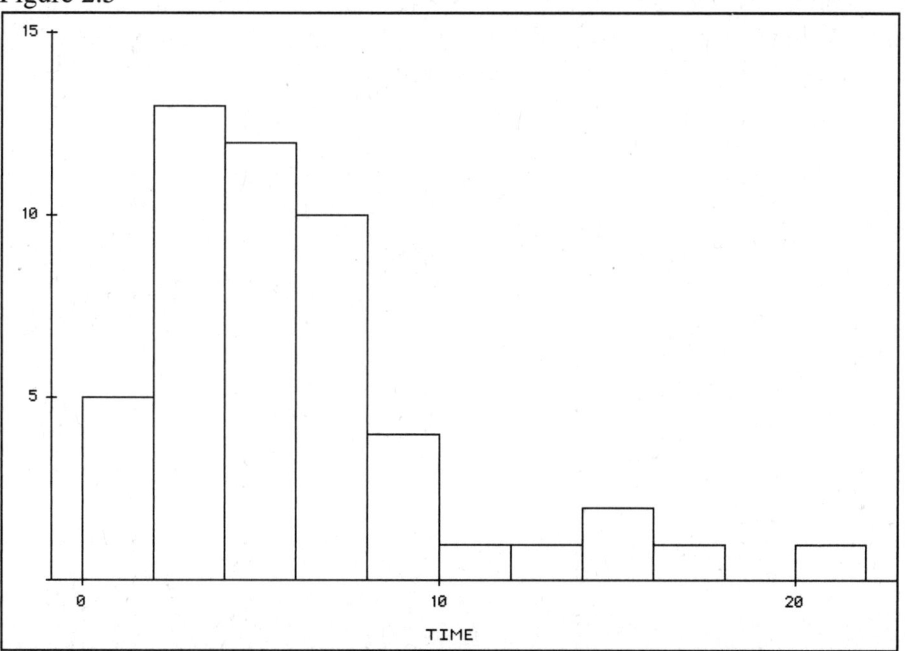

DDXL offers a quick and easy method for creating the histogram shown above. Another option for generating a histogram is to use the Histogram option provided within the Excel program. While more difficult to utilize than the DDXL histograms, Excel offers many more options for creating and histograms. We illustrate this technique and some of the Excel options below.

Once the data is available for analysis, select the **Data** tab in the Ribbon and locate the **Analysis** group. If it has been loaded as an add-in, a Data Analysis option should be available, 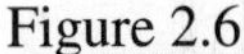

. Click on the Data Analysis to open the Data Analysis menu. From the Data Analysis menu, select the Histogram option (see Figure 2.6). Click OK.

Figure 2.6

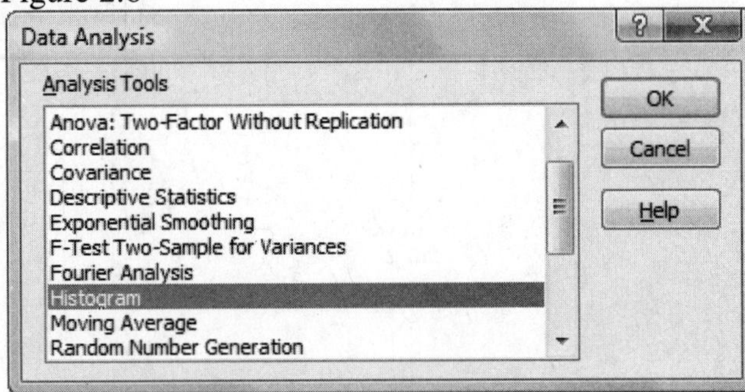

Perform these steps to enter the histogram menu within Excel. There are many options available within this menu. Our purpose here is to demonstrate the easiest method necessary to generate a histogram of the data, and to show the commands necessary to generate a histogram that matches the one shown in the text.

The easiest method to generate a histogram is shown in Figure 2.7. First, enter the rows and columns where the data is located in the **Input Range** of the histogram menu. This can be done by **typing** the location or by **clicking and dragging** over the appropriate data cells in your worksheet. The next step is to specify the **Output Range**. We have chosen to begin the output at column F, row 1 by **typing F1** in the Output Range line of the histogram menu. We have the option to place this output in a new worksheet by specifying the New Worksheet Ply option. Finally, in order to generate the histogram, it is necessary to **check** the **Chart Output** option in the histogram menu. Click **OK**.

Figure 2.7

Histogram
Input
Input Range: $A$2:$A$51
Bin Range:
Labels
Output options
Output Range: $F$1
New Worksheet Ply:
New Workbook
Pareto (sorted histogram)
Cumulative Percentage
Chart Output
OK
Cancel
Help

Excel generates two pieces of information and places this output beginning at the location we specified above. The first is a table of the data that is being charted. The table contains two pieces of information, Bin and Frequency. Bin (see Table 2.2) refers to the upper endpoint of the histogram bar that is to be drawn and Frequency is the number of observations that will be included in the corresponding bar.

Table 2.2

Bin	Frequency
1.25	1
3.961428571	17
6.672857143	16
9.384285714	8
12.09571429	3
14.80714286	3
17.51857143	1
More	1

Together, this information is used by Excel to generate the histogram (see Figure 2.8). The size of the histogram can be altered to make viewing the chart easier. Simply click on the histogram and stretch the squares on the outline of the histogram to make the display larger or smaller.

Figure 2.8

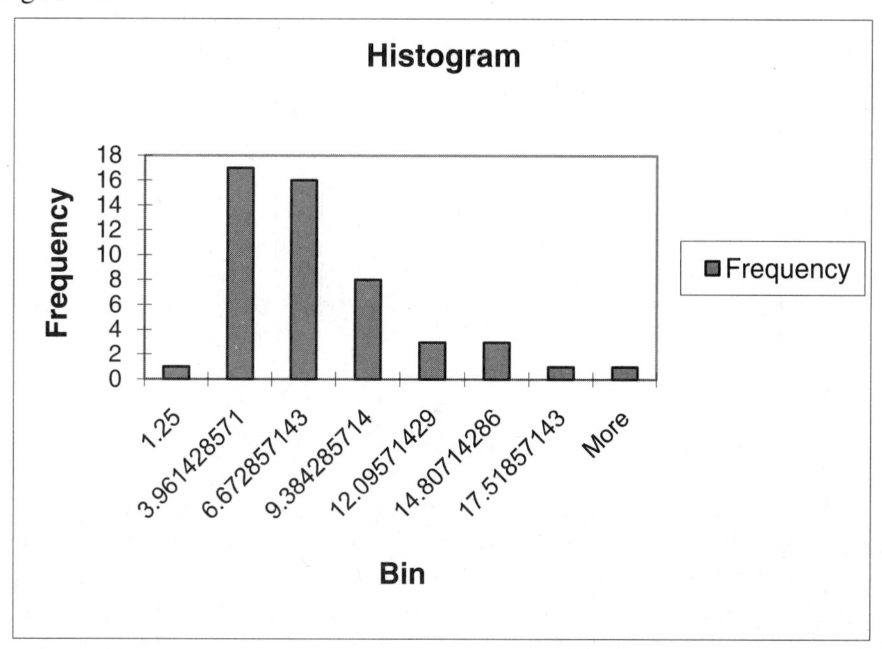

These commands generate a histogram that summarizes the data. Comparing this histogram to the one shown in the text, we see two major differences. First, this histogram has displayed the bars as being separated from one another, while the histogram in the text has bars that touch. The second difference is that the intervals, or interval endpoints, used in the two histograms differ. Both of these differences can be addressed using various options within Excel.

It is important to emphasize that no one histogram that is produced from the data is considered the "correct" one. Our purpose in duplicating the histogram presented in the text is to introduce the user to some of the many options that are available within the histogram menus of Excel. Producing a histogram comparable to the one in the book will allow for easier comparison with the stem-and-leaf display and for easier interpretation of the results.

The touching bars can be adjusted by **right clicking** on any one of the bars generated in the histogram above. Next, select the **Format Data Series** option listed from those given. Click on **Series Options** in the left-hand box and change the **Gap Width** to 0% to assign no gap between bars (see Figure 2.9).

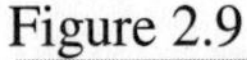

Figure 2.9

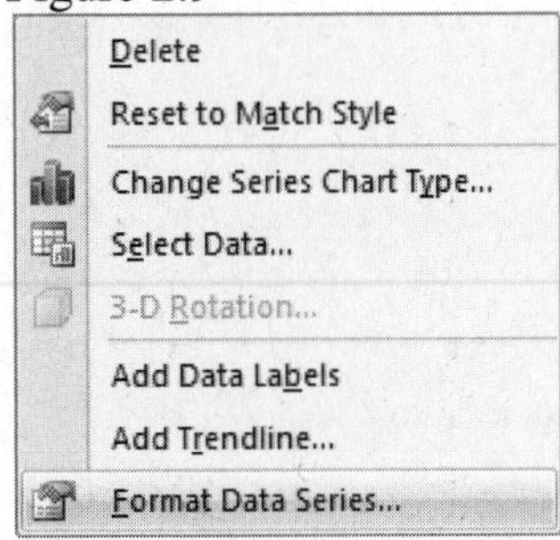

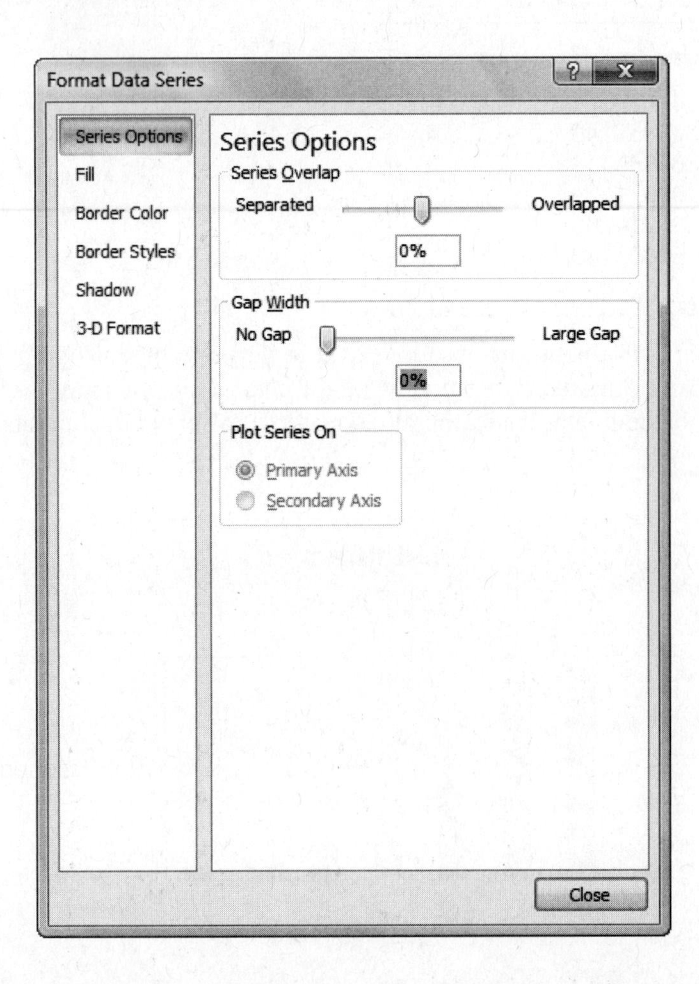

To change the intervals used by Excel requires the addition of a new column in the dataset used for the analysis. You must specify what Bin values that Excel should use to construct the histogram. The Bin values represent the largest endpoint of the bars generated in the chart. To duplicate the histogram presented in the text, with intervals of 2 days, the bar endpoints that we select are shown in Figure 2.10. This column of values must be entered alongside the data and chosen as the Bin Range (see Figures 2.11).

Figure 2.10

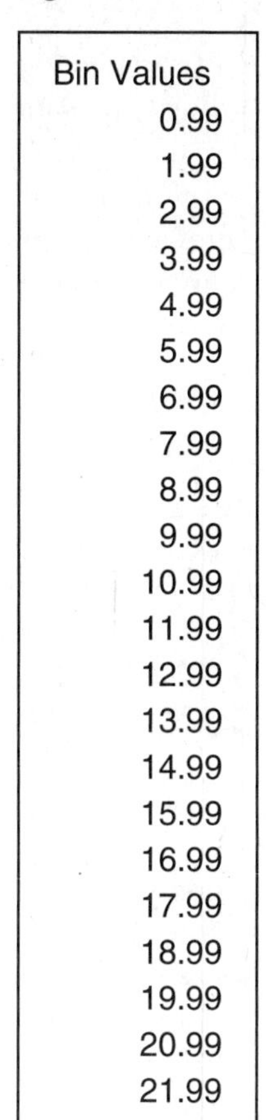

Bin Values
0.99
1.99
2.99
3.99
4.99
5.99
6.99
7.99
8.99
9.99
10.99
11.99
12.99
13.99
14.99
15.99
16.99
17.99
18.99
19.99
20.99
21.99

Figure 2.11

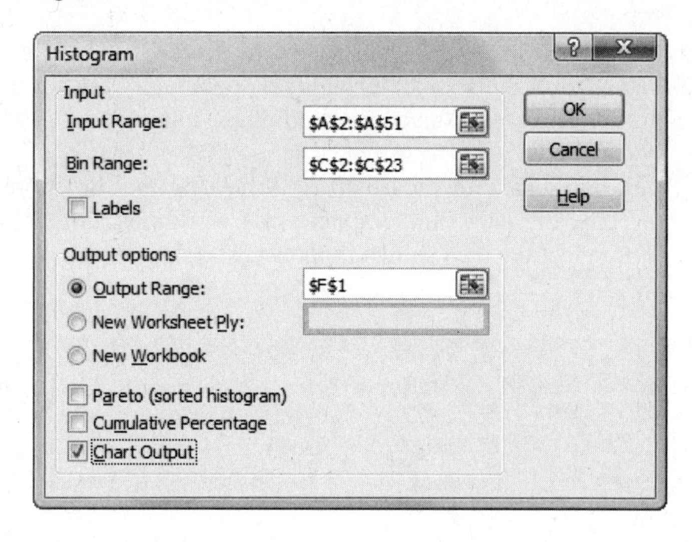

Together, these two changes can be used to produce a histogram that is like the one shown in the text (see Figure 2.12). Be sure to read through Example 2.2 in the text to understand how to interpret these results.

Figure 2.12

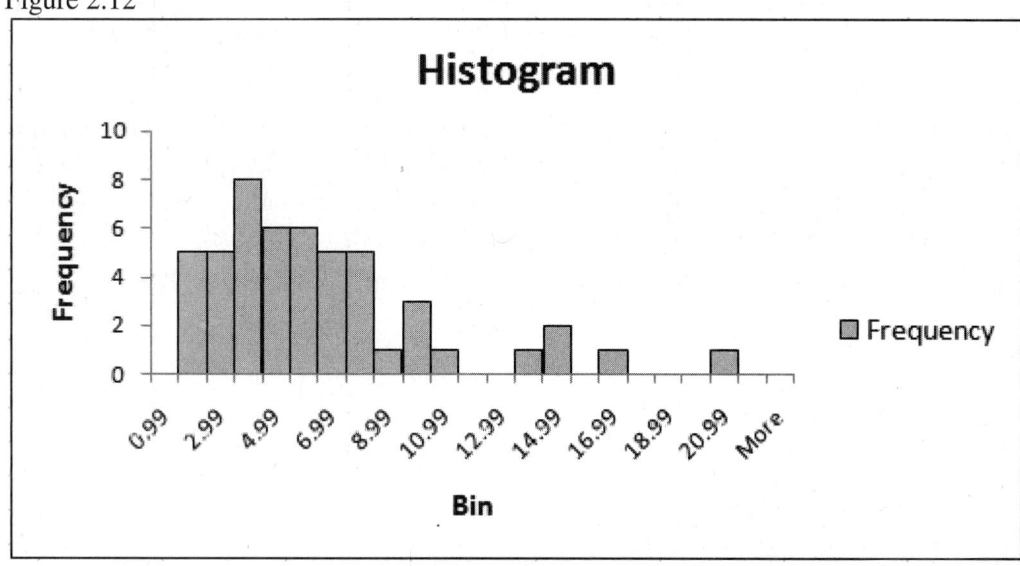

2.2.2 Pie Charts

The second type of graphical technique constructed by Excel is the pie chart. Since none of the chapter examples from *Statistics for Business and Economics* specifically ask for a pie chart, we will use the data from Table 2.1 in the text and create the corresponding pie chart.

Exercise 2.2: **Open** the Data File **Forbes40** by following the directions found in the preface of this manual. If done correctly, the data should appear in a workbook similar to that shown below in Figure 2.13. Use the mouse to select the data shown in the workbook.

Figure 2.13

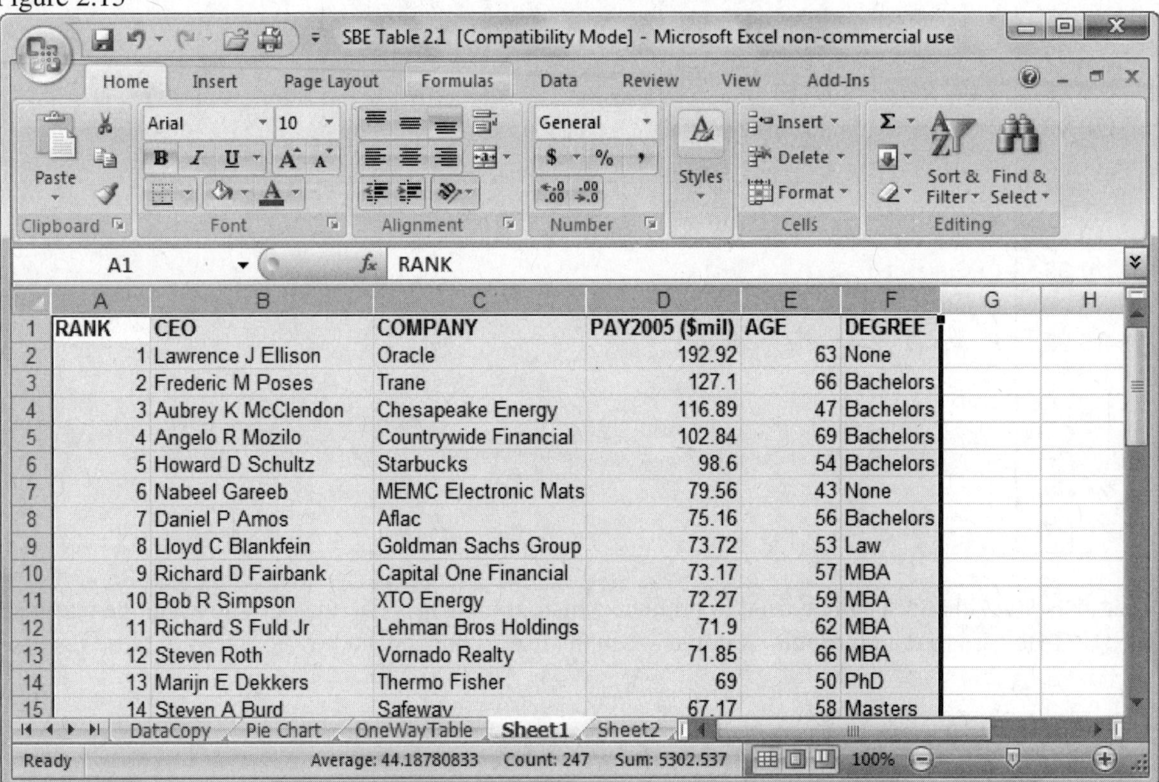

RANK	CEO	COMPANY	PAY2005 ($mil)	AGE	DEGREE
1	Lawrence J Ellison	Oracle	192.92	63	None
2	Frederic M Poses	Trane	127.1	66	Bachelors
3	Aubrey K McClendon	Chesapeake Energy	116.89	47	Bachelors
4	Angelo R Mozilo	Countrywide Financial	102.84	69	Bachelors
5	Howard D Schultz	Starbucks	98.6	54	Bachelors
6	Nabeel Gareeb	MEMC Electronic Mats	79.56	43	None
7	Daniel P Amos	Aflac	75.16	56	Bachelors
8	Lloyd C Blankfein	Goldman Sachs Group	73.72	53	Law
9	Richard D Fairbank	Capital One Financial	73.17	57	MBA
10	Bob R Simpson	XTO Energy	72.27	59	MBA
11	Richard S Fuld Jr	Lehman Bros Holdings	71.9	62	MBA
12	Steven Roth	Vornado Realty	71.85	66	MBA
13	Marijn E Dekkers	Thermo Fisher	69	50	PhD
14	Steven A Burd	Safeway	67.17	58	Masters

Suppose we wish to create a bar graph for the Degree variable in the data set. Click on the **DDXL Add-In** menu. Click on the **Charts and Plots** option to access the **Charts and Plots Dialog** menu (see Figure 2.14 below). Click on the ▼ in the Function Type box to access the different charts available to select. Highlight the **Pie Chart** option. Highlight the DEGREE variable in the Names and Columns Box. Click on the ◀ to the left of the Categorical Variable box to create a pie chart for the Degree variable. Click **OK**. The pie chart is shown in Figure 2.15.

Figure 2.14

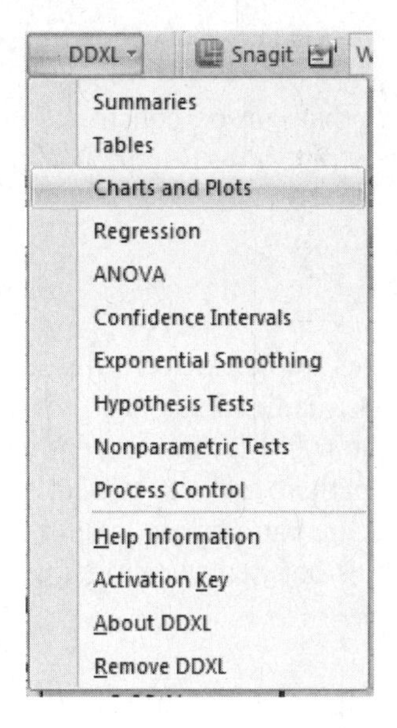
DDXL
Snagit
Summaries
Tables
Charts and Plots
Regression
ANOVA
Confidence Intervals
Exponential Smoothing
Hypothesis Tests
Nonparametric Tests
Process Control
Help Information
Activation Key
About DDXL
Remove DDXL

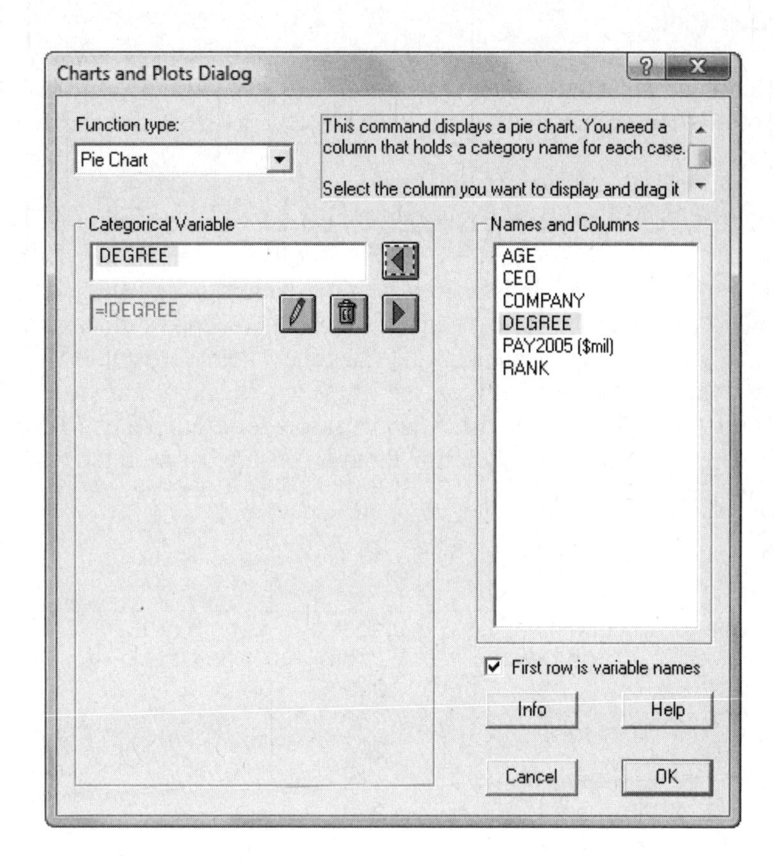
Charts and Plots Dialog
Function type:
Pie Chart
This command displays a pie chart. You need a column that holds a category name for each case.
Select the column you want to display and drag it
Categorical Variable
DEGREE
=!DEGREE
Names and Columns
AGE
CEO
COMPANY
DEGREE
PAY2005 ($mil)
RANK
First row is variable names
Info
Help
Cancel
OK

Figure 2.15

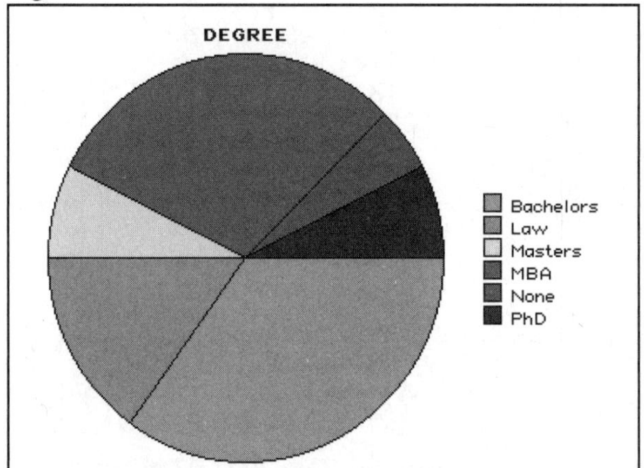
DEGREE
Bachelors
Law
Masters
MBA
None
PhD

2.2.3 Scatter Plots

Another graphing technique discussed in *Statistics for Business and Economics* that can be constructed within Excel is the scatter plot. We use Example 2.19 from the text to demonstrate how to conduct a scatter plot.

Exercise 2.3: *Statistics for Business and Economics* Example 2.19.

A medical item used to administer to a hospital patient is called a **factor**. For example, factors can be intravenous (IV) tubing, IV fluid, needles, shave kits, bedpans, diapers, dressings, medications, and even code carts. The coronary care unit at Bayonet Point Hospital (St. Petersburg, Florida) recently investigated the relationship between the number of factors administered per patient and the patient's length of stay (in days). Data on these two variables for a sample of 50 coronary care patients are given in Table 2.3. Use a scattergram to describe the relationship between the two variables of interest, number of factors, and length of stay.

Table 2.3

Number of Factors	Length of Stay (in Days)	Number of Factors	Length of Stay (in Days)	Number of Factors	Length of Stay (in Days)
231	9	233	8	115	4
323	7	260	4	202	6
113	8	224	7	206	5
208	5	472	12	360	6
162	4	220	8	84	3
117	4	383	6	331	9
159	6	301	9	302	7
169	9	262	7	60	2
55	6	354	11	110	2
77	3	142	7	131	5
103	4	286	9	364	4
147	6	341	10	180	7
230	6	201	5	134	6
78	3	158	11	401	15
525	9	243	6	155	4
121	7	156	6	338	8
248	5	184	7		

Solution:

Before we begin, we must access the data set for this example. **Open** the Data File **Medfactors** by following the directions found in the preface of this manual. Use the mouse to select the data shown in the workbook. We need to create a scattergram for the two variables given in the data set. Click on the **DDXL Add-In** menu. Click on the **Charts and Plots** option to access the **Charts and Plots Dialog** menu (see Figure 2.16 below). Click on the in the Function Type box to access the different charts available to select. Highlight the **Scatter Plot** option. Highlight the FACTORS variable from the Names and Columns box and click on the to the left of the x-Axis Variable box to plot the Factors variable on the horizontal axis. Highlight the LOS variable from the Names and Columns box and click on the to the left of the y-Axis Variable box to plot the Length of Stay variable on the vertical axis. Click **OK** to create the scatter plot shown in Figure 2.17 below.

Figure 2.16

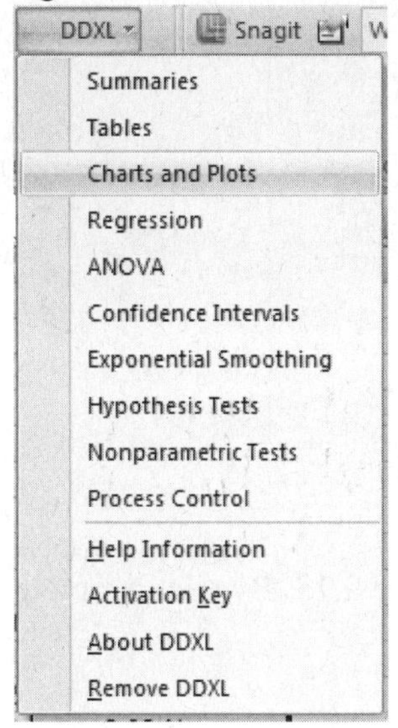

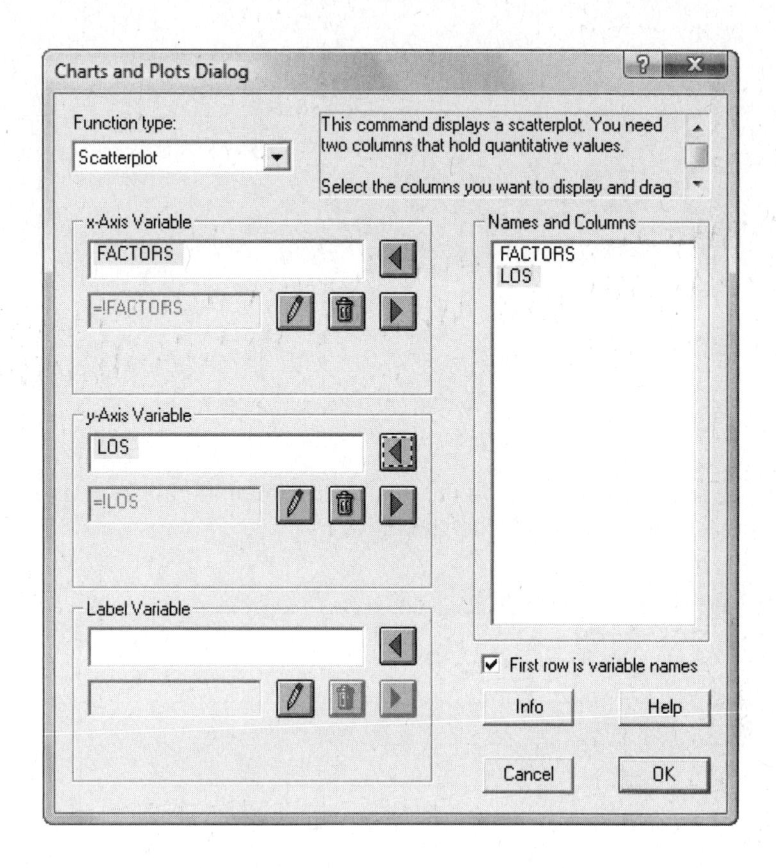

Figure 2.17

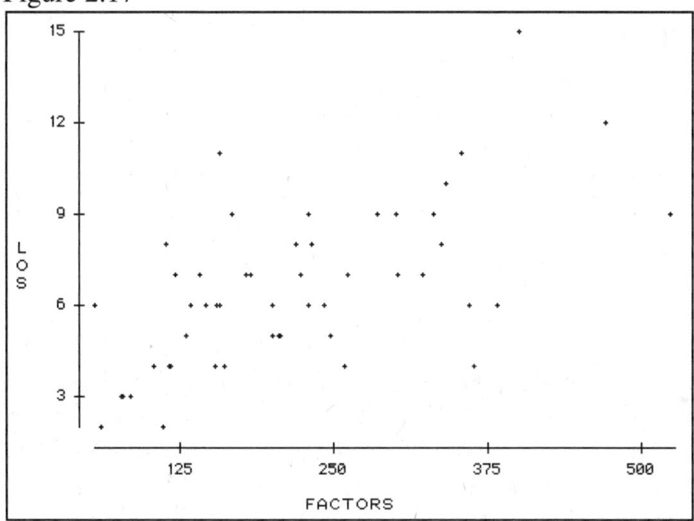

DDXL offers a quick and easy method for creating the scatter plot shown above. Another option for generating a scatter plot is to use the Charting options provided within the Excel program. While more difficult to utilize than the DDXL scatter plots, Excel offers many more options for creating these plots. We illustrate this technique below.

Excel offers a wide of charts and graphs using the Charts group located in the Insert tab on the Ribbon. By clicking on the Scatter plot option, we get to choose the type of scatter plot to create.

Figure 2.18

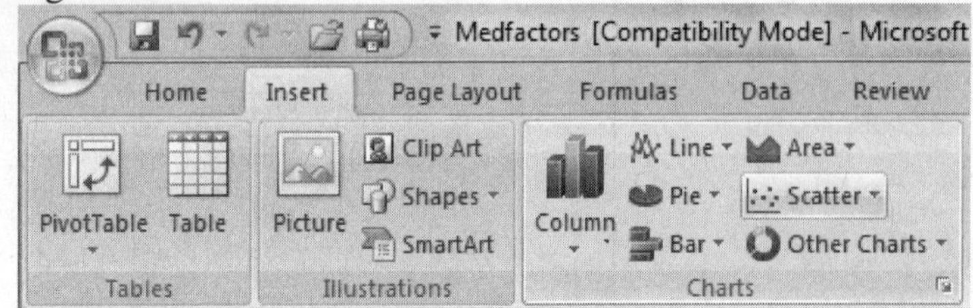

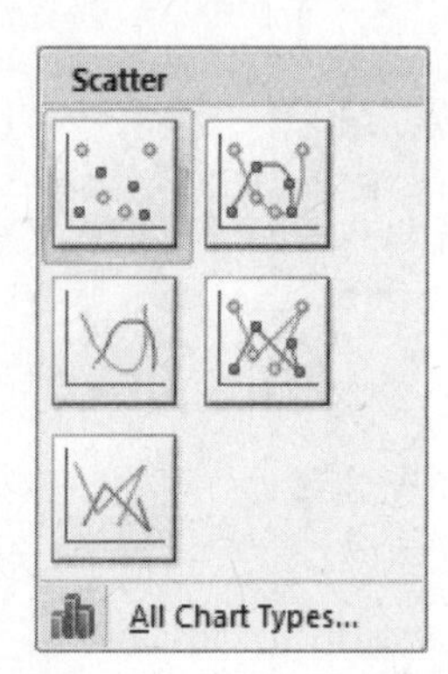

The basic scatter plot we see in the text is the option highlighted above in Figure 2.18. By clicking on it, Excel creates the scatter plot shown below.

Figure 2.19

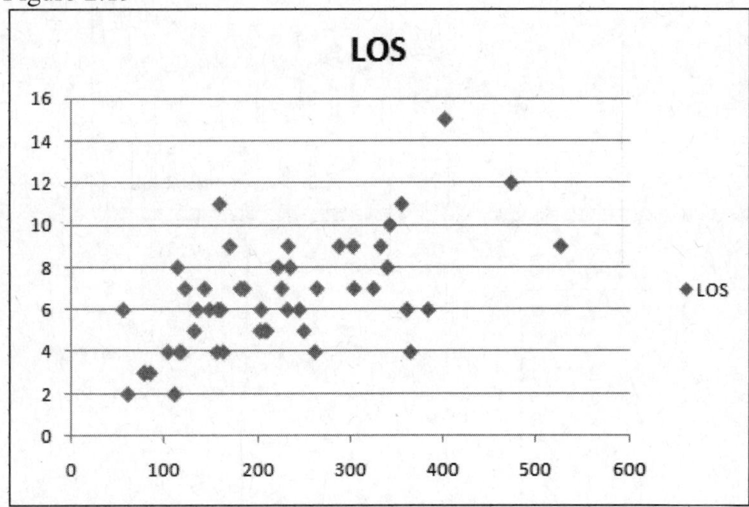

Excel offers many different options when creating scatter plots and other charts. Review the help features within Excel to learn more about these options.

One final note before we leave the scatter plot options within Excel. The scatter plot can also be used to create the time series plot discussed in *Statistics for Business and Economics*. When creating the scatter plot, the measure of time over which the data was collected should be used as the x-axis variable in the time series plot. All other steps are identical to that of the scatter plot discussed here.

2.2.4 Box Plots

DDXL allow the user to easily create box plots for quantitative data. We demonstrate by using the data from Example 2.2 of the *Statistics for Business* and Economics text. A box plot for the 50 processing times is constructed below.

Exercise 2.4: Use DDXL to create a box plot for the 50 processing times of the data found in Example 2.2 in the text (Pricequotes data set).

Solution:

Before we begin, we must access the data set for this example. **Open** the Data File **Pricequotes** by following the directions found in the preface of this manual. Use the mouse to select the data shown in the workbook. We wish to create a box plot for the processing time variables given in the data set. Click on the **DDXL Add-In** menu. Click on the **Charts and Plots** option to access the **Charts and Plots Dialog** menu (see Figure 2.20 below). Click on the ▾ in the Function Type box to access the different charts available to select. Highlight the **Box Plot** option. Highlight the TIME variable found in the Names and Columns box and click on the ◀ to the left of the Quantitative Variable box to create a box plot for the Time variable. Click **OK** to create the box plot shown in Figure 2.20 below.

Figure 2.20

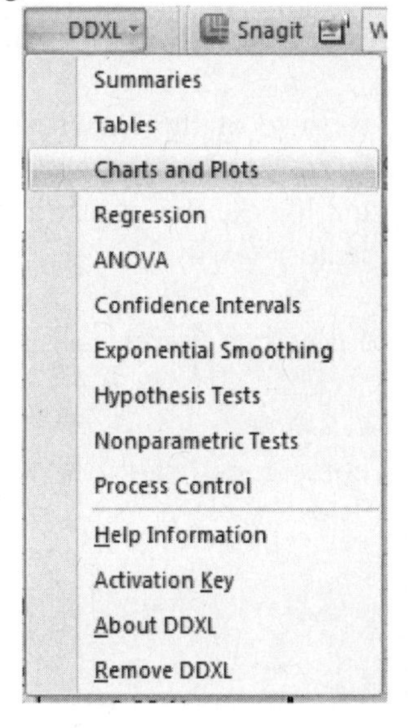

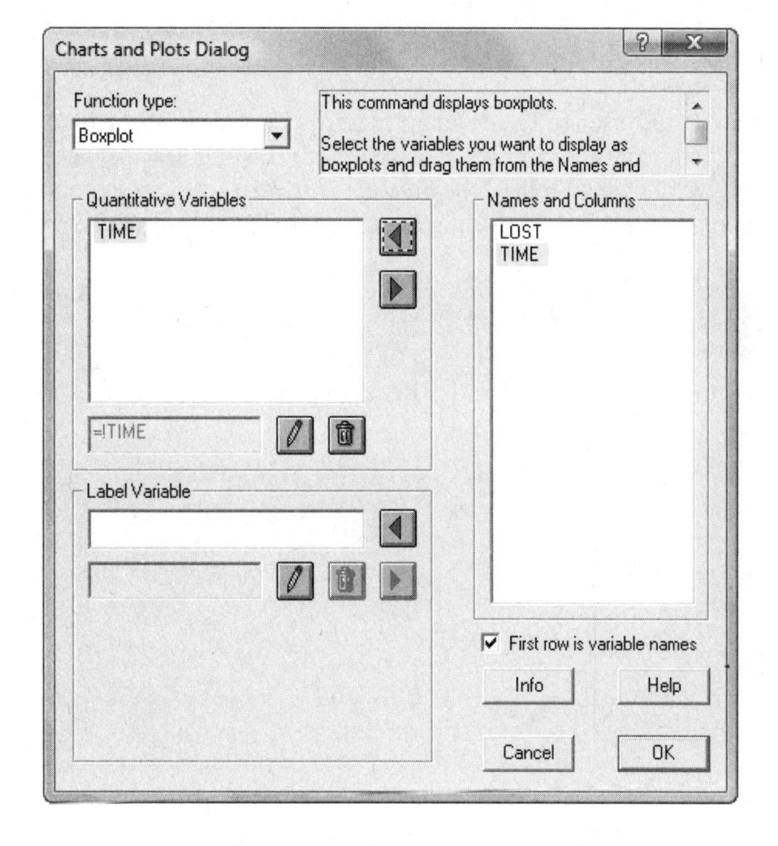

Figure 2.21

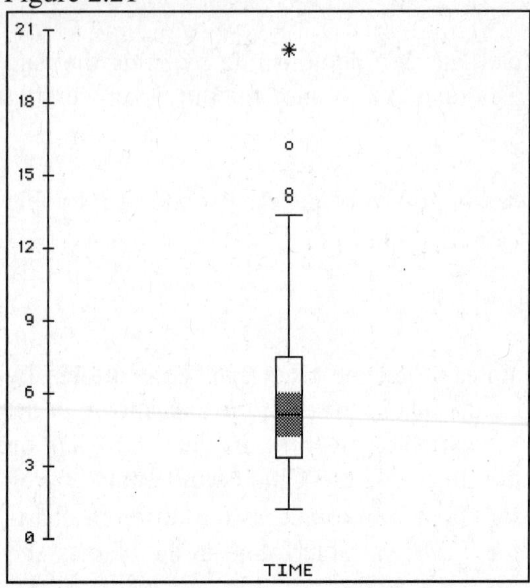

2.3 Numerical Techniques in Excel

2.3.1 Measures of Center

Excel allows the user to create many descriptive measures of data through the use of the Descriptive Statistics data analysis. While Excel doesn't distinguish between the different types of numerical measures, we choose to follow the *Statistics for Business and Economics* text and look at the measures of center, spread, and relative standing one at a time. We begin with measures of center.

Exercise 2.5 We use *Statistics for Business and Economics* Example 2.4 to illustrate the measures of center.

Calculate the sample mean for the R&D expenditure percentages of the 50 companies listed below in Table 2.4.

Table 2.4

Company	Percentage	Company	Percentage	Company	Percentage
1	13.5	18	6.9	35	8.5
2	8.4	19	7.5	36	9.4
3	10.5	20	11.1	37	10.5
4	9	21	8.2	38	6.9
5	9.2	22	8	39	6.5
6	9.7	23	7.7	40	7.5
7	6.6	24	7.4	41	7.1
8	10.6	25	6.5	42	13.2
9	10.1	26	9.5	43	7.7
10	7.1	27	8.2	44	5.9
11	8	28	6.9	45	5.2
12	7.9	29	7.2	46	5.6
13	6.8	30	8.2	47	11.7
14	9.5	31	9.6	48	6
15	8.1	32	7.2	49	7.8
16	13.5	33	8.8	50	6.5
17	9.9	34	11.3		

Solution:

Before we begin, we must access the data set for this example. **Open** the Data File **R&D** by following the directions found in the preface of this manual. Once the data is available, click on the **Data** tab and choose the **Data Analysis** option found in the **Analysis** group. Next, highlight the **Descriptive Statistics** option (see Figure 2.22) and click **OK**.

Figure 2.22

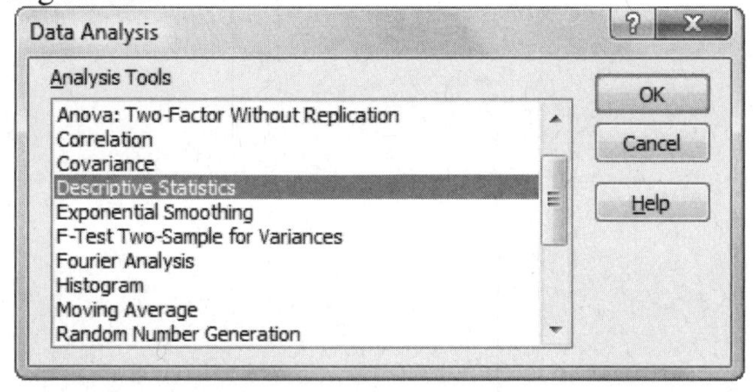

From the Descriptive Statistics menu, the user must specify the **Input Range**, the **Output Range**, and which statistics are desired. The Input Range indicates the worksheet location of the data to be analyzed. Either type or highlight with the mouse and enter the data set location for the Input Range (see Figure 2.23). The Output Range can either be a location within the current worksheet or a new Worksheet that you define. We opt to place the output in cell D1 of the current worksheet. Finally, the **Summary Statistics** box needs to be **checked** to generate the measures of center that are desired. Click **OK**.

Figure 2.23

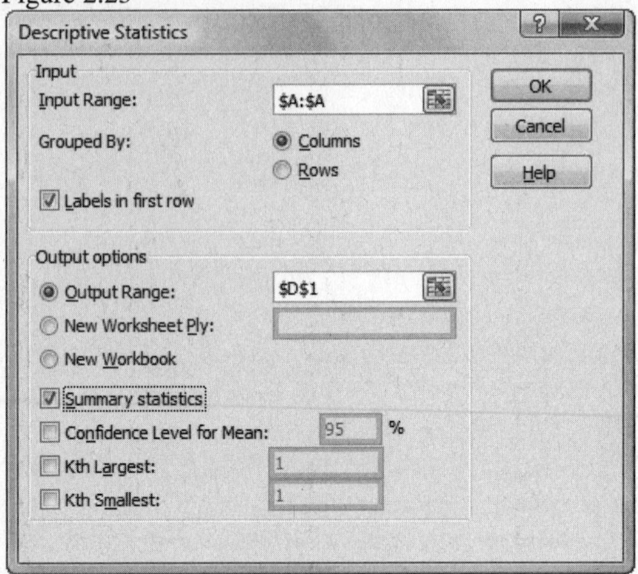

Excel calculates the three measures of center, mean, median, and mode for the data set of interest (see Table 2.5). The mean R&D expenditure for the 50 companies is reported to be 8.492 percent.

Table 2.5

RDPct	
Mean	8.492
Standard Error	0.2800997
Median	8.05
Mode	6.9
Standard Deviation	1.9806039
Sample Variance	3.9227918
Kurtosis	0.4192877
Skewness	0.8546013
Range	8.3
Minimum	5.2
Maximum	13.5
Sum	424.6
Count	50

Calculating the measures of center for other data sets requires only changing the Input Range values in the Descriptive Statistics menu above. Notice that the values for both the sample median and sample mode are also given on this printout.

2.3.2 Measures of Spread

The three measures of spread, range, standard deviation, and variance are found in the same manner as the measures of center above. We will use Example 2.10 from *Statistics for Business and Economics* to demonstrate.

Exercise 2.6 We use *Statistics for Business and Economics* Example 2.10.

Use the computer to find the sample variance s^2 and the sample standard deviation s for the 50 companies' percentage of revenues spent on R & D. The data is shown in Table 2.6.

Table 2.6

Company	Percentage	Company	Percentage	Company	Percentage
1	13.5	18	6.9	35	8.5
2	8.4	19	7.5	36	9.4
3	10.5	20	11.1	37	10.5
4	9	21	8.2	38	6.9
5	9.2	22	8	39	6.5
6	9.7	23	7.7	40	7.5
7	6.6	24	7.4	41	7.1
8	10.6	25	6.5	42	13.2
9	10.1	26	9.5	43	7.7
10	7.1	27	8.2	44	5.9
11	8	28	6.9	45	5.2
12	7.9	29	7.2	46	5.6
13	6.8	30	8.2	47	11.7
14	9.5	31	9.6	48	6
15	8.1	32	7.2	49	7.8
16	13.5	33	8.8	50	6.5
17	9.9	34	11.3		

Solution:

Before we begin, we must access the data set for this example. **Open** the Data File **R&D** by following the directions found in the preface of this manual. Once the data is available, click on the **Data** tab and choose the **Data Analysis** option found in the **Analysis** group. Next, highlight the **Descriptive Statistics** option (see Figure 2.22 above) and click **OK**.

From the Descriptive Statistics menu, the user must specify the **Input Range**, the **Output Range**, and which statistics are desired. The Input Range indicates the worksheet location of the data to be analyzed. Either type or highlight with the mouse and enter the data set location for the Input Range (see Figure 2.23 above). The Output Range can either be a location within the current worksheet or a new Worksheet that you define. We opt to place the output in cell D1 of the current worksheet. Finally, the **Summary Statistics** box needs to be **checked** to generate the measures of variation that are desired. Click **OK**.

Excel calculates three measures of spread (range, standard deviation, and variance) for the data set of interest (see Table 2.7). The sample variance for the R&D expenditure of the 50 companies is reported to be 3.922792 and the sample standard deviation is 1.980604 percent.

Table 2.7

RDPct	
Mean	8.492
Standard Error	0.2800997
Median	8.05
Mode	6.9
Standard Deviation	1.9806039
Sample Variance	3.9227918
Kurtosis	0.4192877
Skewness	0.8546013
Range	8.3
Minimum	5.2
Maximum	13.5
Sum	424.6
Count	50

Calculating the measures of spread for other data sets requires only changing the Input Range values in the Descriptive Statistics menu above.

2.3.3 Measure of Relative Standing

Excel allows the user to calculate the two measures of relative standing, percentiles and z-scores through the use of two of its many functions. We first look at how Excel calculates percentiles.

Exercise 2.7: Use the data from Example 2.14 to calculate the 25th and 90th percentile of the R&D percentages.

Solution:

Before we begin, we must access the data set for this example. **Open** the Data File **R&D** by following the directions found in the preface of this manual. Once the data is available, click the Σ ▾ icon located in the **Editing** group of the **Home** tab.

Figure 2.24

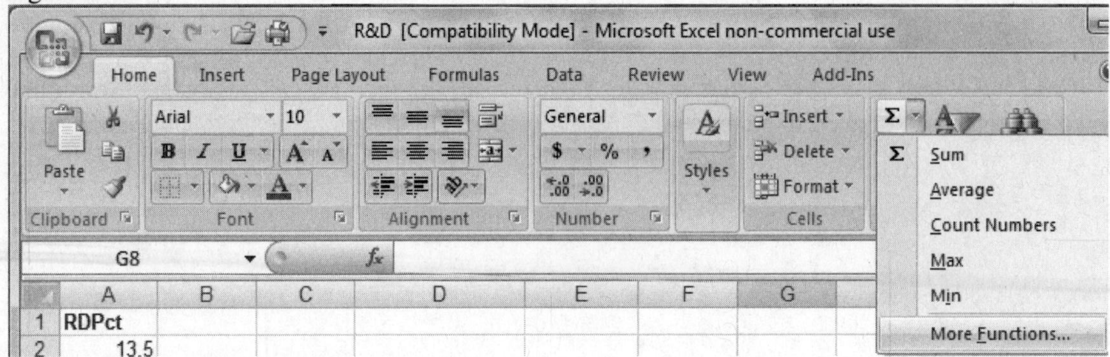

By clicking on the down arrow, you can access the **More Functions** menu. Choose the **Statistical Function Category** and cursor down until you reach the function name **PERCENTILE** (see Figure 2.25). The PERCENTILE function has the form:

PERCENTILE(array,k)

where **array** represents the location of the data set that you want to find the percentile for, and **k** is a number between 0 and 1 that represents the percentile that is desired.

Figure 2.25

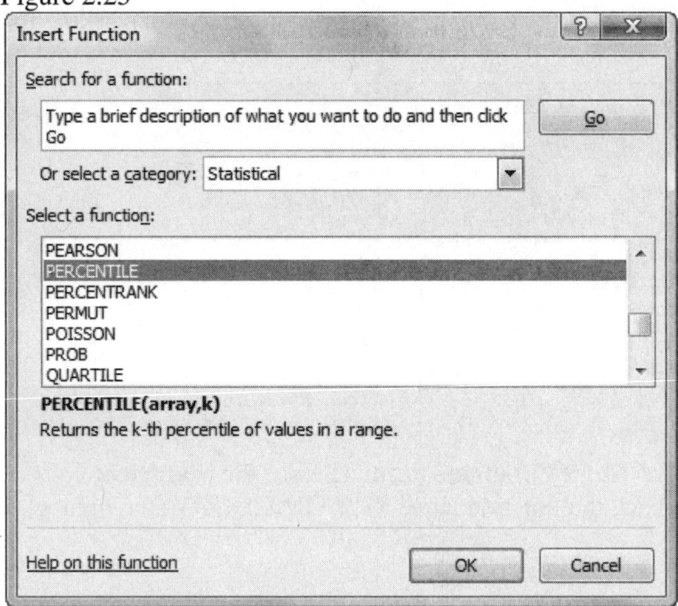

For this example, the 50 R&D percentages are located in column A in rows 2 through 51. We assign the **Array** location to be A2:A51 (see Figure 2.26). We also assign the value of **K** to be .25 representing the 25th percentile. Click **OK**.

Figure 2.26

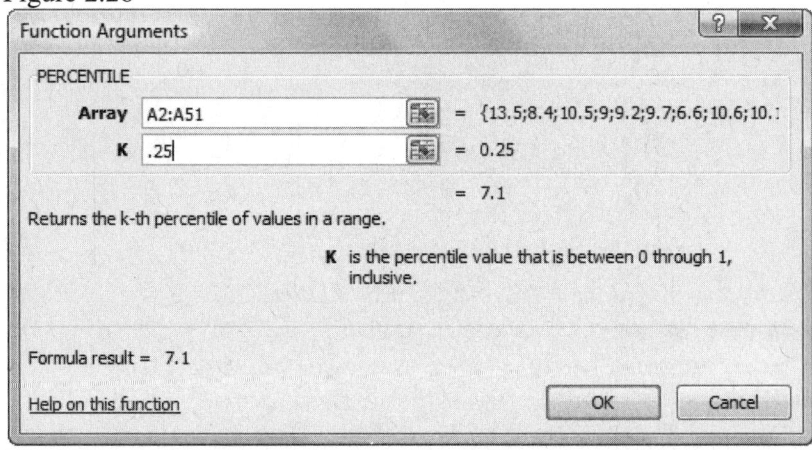

Excel returns a value of 7.1. We interpret that 7.1% represents the 25th percentile of the 50 R&D percentages in our data set. By changing the K to .95, Excel tells us that the 95th percentile is the value 12.525%. By changing the value of K, we can find any percentile we want.

The second measure of relative standing is the z-score. Again, we turn in Excel to a function that will allow the user to calculate values for a z-score. For purposes of illustration, we will again use the data from Example 2.10 to find a z-value.

Exercise 2.8: Use the 50 R&D percentages to find the z-score for an R&D percentage of 10%.

Solution:

Before we begin, we must access the data set for this example. **Open** the Data File **R&D** by following the directions found in the preface of this manual. Once the data is available, click the 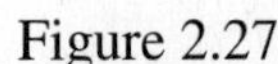 icon located in the **Editing** group of the **Home** tab.

Figure 2.27

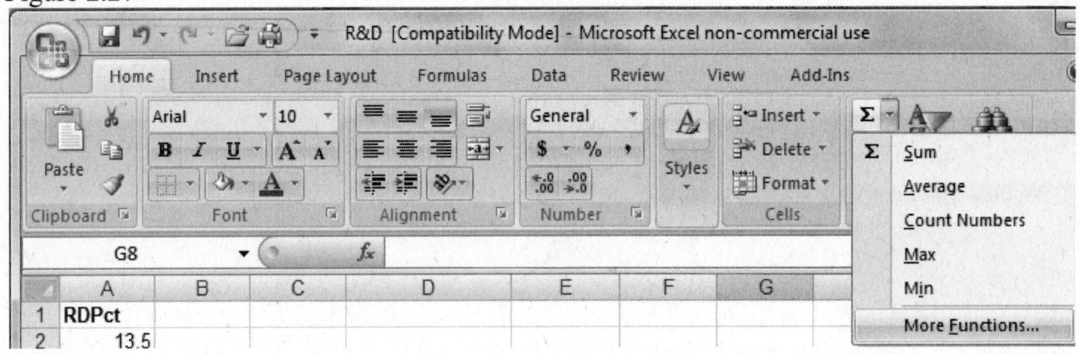

By clicking on the down arrow, you can access the **More Functions** menu. Choose the **Statistical Function Category** and cursor down until you reach the function name **STANDARDIZW** (see Figure 2.28). The PERCENTILE function has the form:

STANDARDIZE (x, mean, standard_dev)

where **x** represents the value that you wish to determine the z-score for,
mean represents the mean of the data set that you want to find the z-score for, and
standard deviation represents the standard deviation of the data set that you want to find the z-score for.

Figure 2.28

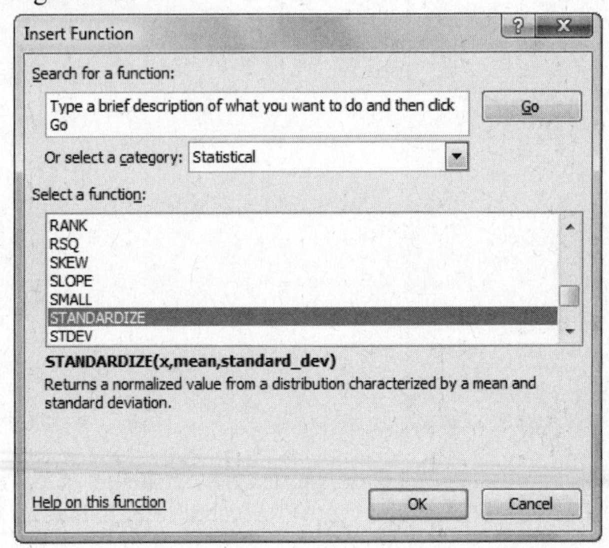

For this example, we use the value of 10 as our choice for **X** in the STANDARDIZE function (see Figure 2.28). From the work we did on Exercise 2.7, we know to use a value of 8.492 for the **mean** and a value of 1.980604 for the **standard deviation**. Click **OK**.

Figure 2.28

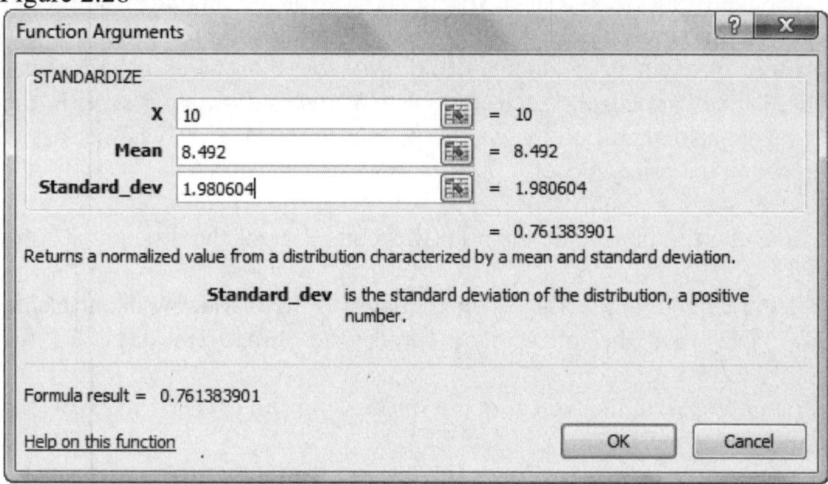

Excel returns a value of 0.761383901. We make the interpretation that an R&D percentage of 10% would fall approximately .76 standard deviation above the mean R&D percentage of the 50 companies. By changing the values of X, Mean, and Standard Deviation, we can find z-scores for a wide variety of situations.

Technology Lab

2.150 A manufacturer of industrial wheels is losing many profitable orders because of the long time it takes the firm's marketing, engineering, and accounting departments to develop price quotes for potential customers. To remedy this problem the firm's management would like to set guidelines for the length of time each department should spend developing price quotes. To help develop these guidelines, 50 requests for price quotes were randomly selected from the set of all price quotes made last year; the processing time was determined for each department. These times are displayed in the table below and are contained in the Excel file **Lostquotes**. The price quotes are also classified by whether they were "lost" (i.e., whether or not the customer placed an order after receiving the price quote).

a. Construct a bar graph and a pie chart to determine the relative frequencies of the orders that were lost.
b. Construct a histogram for the price quote processing times (in days) of the marketing department.
c. Construct a stem-and-leaf display for the price quote processing times (in days) of the engineering department.
d. Construct a scatter plot that compares the price quote processing times (in days) of the engineering and the accounting departments.
e. Construct a box plot for the price quote processing times (in days) of the marketing department.
f. Find the descriptive statistics for the price quote processing times (in days) of the all three departments.

Use the Excel output provided below to check your work.

PRICE QUOTE PROCESSING TIME (IN Days)

Request Number	Marketing	Engineering	Accounting	Lost?	Request Number	Marketing	Engineering	Accounting	Lost?
1	7	6.2	0.1	2	26	0.6	2.2	0.5	2
2	0.4	5.2	0.1	2	27	6	1.8	0.2	2
3	2.4	4.6	0.6	2	28	5.8	0.6	0.5	2
4	6.2	13	0.8	1	29	7.8	7.2	2.2	1
5	4.7	0.9	0.5	2	30	3.2	6.9	0.1	2
6	1.3	0.4	0.1	2	31	11	1.7	3.3	2
7	7.3	6.1	0.1	2	32	6.2	1.3	2	2
8	5.6	3.6	3.8	2	33	6.9	6	10.5	1
9	5.5	9.6	0.5	2	34	5.4	0.4	8.4	2
10	5.3	4.8	0.8	2	35	6	7.9	0.4	2
11	6	2.6	0.1	2	36	4	1.8	18.2	1
12	2.6	11.3	1	2	37	4.5	1.3	0.3	2
13	2	0.6	0.8	2	38	2.2	4.8	0.4	2
14	0.4	12.2	1	2	39	3.5	7.2	7	1
15	8.7	2.2	3.7	2	40	0.1	0.9	14.4	2
16	4.7	9.6	0.1	2	41	2.9	7.7	5.8	2
17	6.9	12.3	0.2	1	42	5.4	3.8	0.3	2
18	0.2	4.2	0.3	2	43	6.7	1.3	0.1	2
19	5.5	3.5	0.4	2	44	2	6.3	9.9	1
20	2.9	5.3	22	2	45	0.1	12	3.2	2
21	5.9	7.3	1.7	2	46	6.4	1.3	6.2	2
22	6.2	4.4	0.1	2	47	4	2.4	13.5	1
23	4.1	2.1	30	1	48	10	5.3	0.1	2
24	5.8	0.6	0.1	2	49	8	14.4	1.9	1
25	5	3.1	2.3	2	50	7	10	2	2

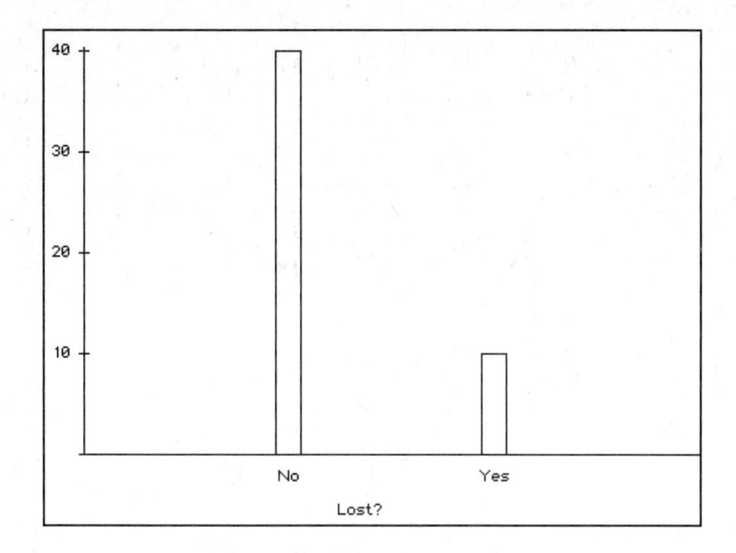
40
30
20
10
No
Yes
Lost?

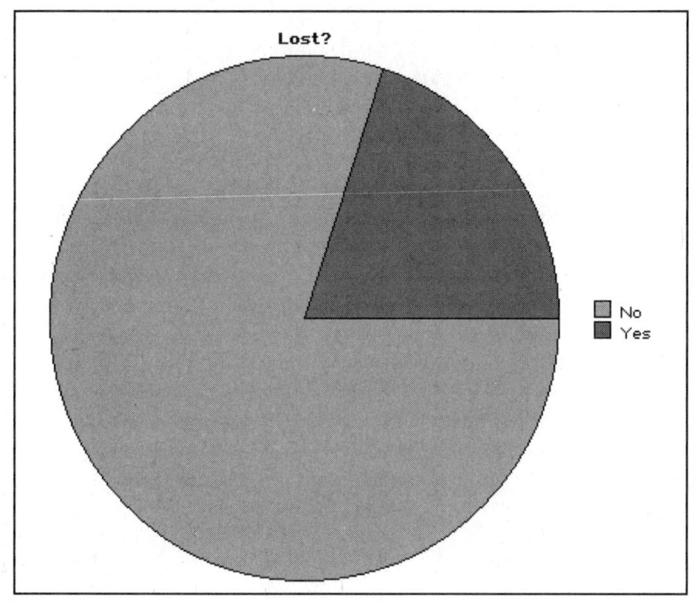
Lost?
No
Yes

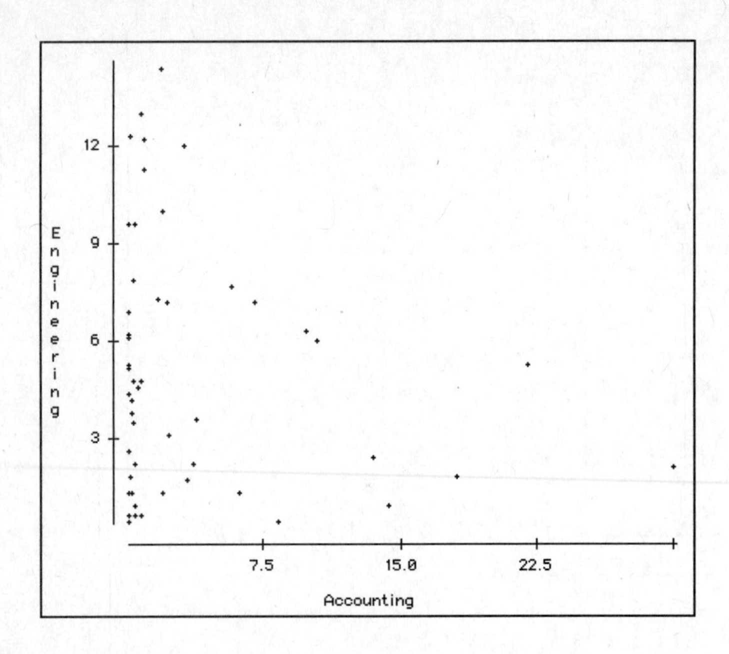

Engineering
12
9
6
3
7.5
15.0
22.5
Accounting

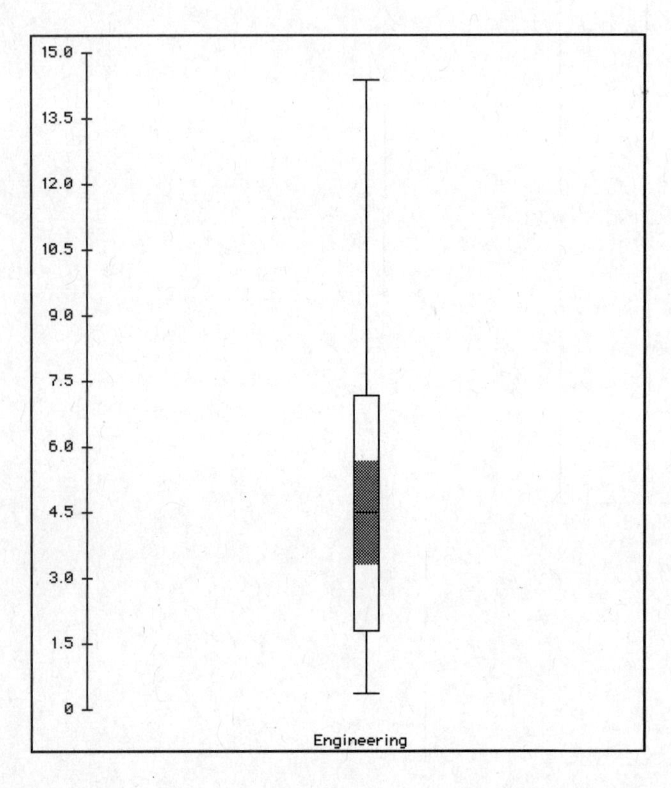

15.0
13.5
12.0
10.5
9.0
7.5
6.0
4.5
3.0
1.5
0
Engineering

	Marketing	*Engineering*	*Accounting*
Mean	4.766	5.044	3.652
Standard Error	0.365431	0.542294	0.884793
Median	5.4	4.5	0.8
Mode	6.2	1.3	0.1
Standard Deviation	2.583986	3.834599	6.256433
Sample Variance	6.676984	14.70415	39.14296
Kurtosis	-0.25619	-0.33934	6.906608
Skewness	-0.10272	0.756246	2.552141
Range	10.9	14	29.9
Minimum	0.1	0.4	0.1
Maximum	11	14.4	30
Sum	238.3	252.2	182.6
Count	50	50	50

Chapter 3
Probability

3.1 Introduction

Chapter 3 introduces the topic of probability and random sampling to the reader. DDXL does not offer any analyses directly related to probability but Excel has two features that we will briefly mention to help users take random samples from populations. The RANDBETWEEN function and the Random Number Generation tool allow the user two methods of drawing random selections from a population of N items. Both procedures can be used to generate random samples and are discussed below.

3.2 Random Sampling

3.2.1 The RANDBETWEEN Function

Section 3.7 in the text defines to the reader what a random sample is and gives a method of generating a random sample that utilizes the random number table found in Table 1 of the *Statistics for Business and Economics* text. The RANDBETWEEN function offered in Excel provides a simpler method of drawing a random sample from a population of known population size. We illustrate this function using Example 3.22 found in the *Statistics for Business and Economics* text.

Exercise 3.1: Suppose you wish to randomly sample five households from a population of 100,000 households to participate in a study.

a. How many different samples can be selected?
b. Use a random number generator to select a random sample.

We are shown in the text that there are $8.33X10^{22}$ possible random samples of five homes that can be selected from a population of 100,000 homes. In part b we are asked to take one of these samples.

We begin by opening up Excel and placing the cursor on any cell in the blank worksheet. We click on the **Home** tab and then select the arrow next to the **Function** Icon in the **Editing** group of options shown below in Figure 3.1.

Figure 3.1

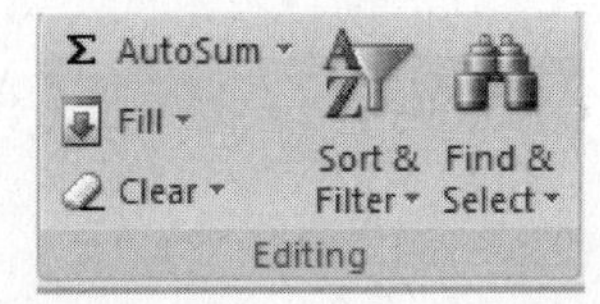

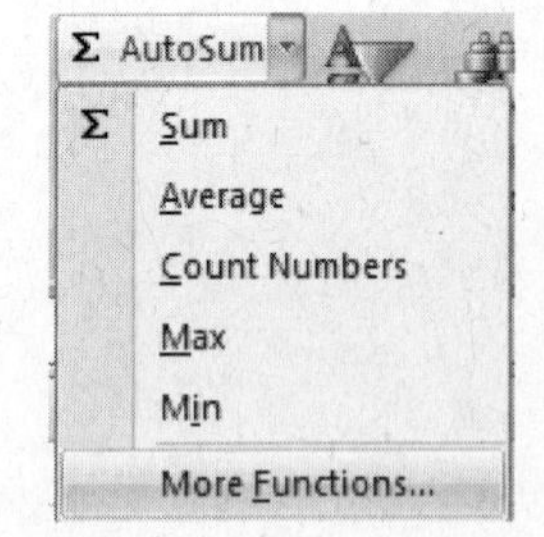

We then select the **More Functions** option to get the **Insert Function** menu shown below in Figure 3.2 Click on the arrow to select that **All** categories are being used and scroll down until you reach the **RANDBETWEEN** function. Click OK to open the **RANDBETWEEN** function.

Figure 3.2

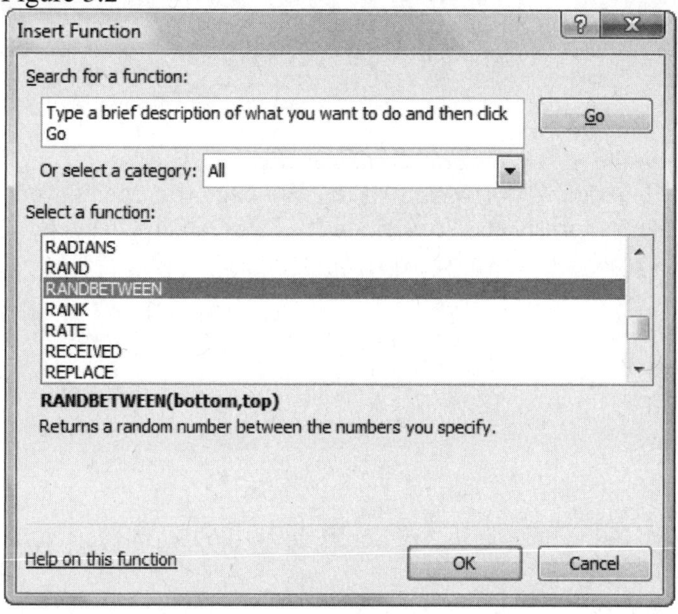

The user is asked to furnish the two values, called **Bottom** and **Top**, that represent the smallest and largest values that we want to sample from. In our example, we wish to sample five homes from within a group of 100,000 homes. We specify a Bottom value of 1 and a Top value of 100,000 in Figure 3.3 below. Click OK. Excel returns a random integer between the values specified. In our example, Excel returned the value of 48,051 (see Figure 3.4). We would sample the home that was labeled as 48,051 as our first home in the sample.

Figure 3.3

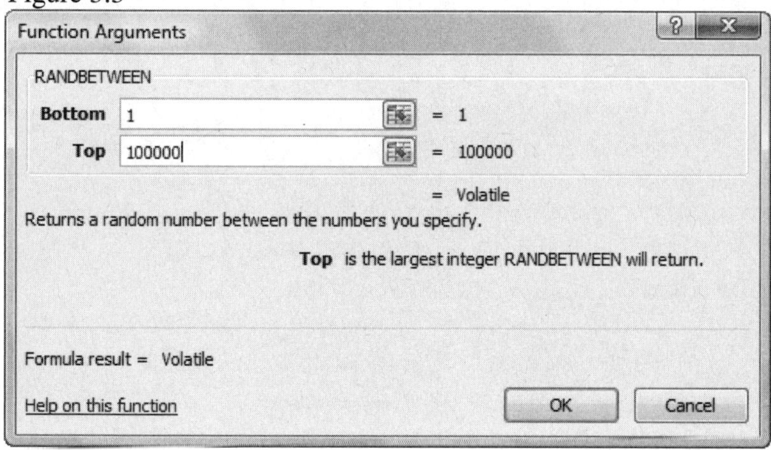

Figure 3.4

A1 f_x =RANDBETWEEN(1,100000)

	A	B	C	D	E	F	G
1	48051						
2							

Note that both the Bottom and Top numbers are possible for selection when using the RANDBETWEEN function. In our example, we would need to repeat this process four additional times to generate a random sample of five homes. The RANDBETWEEN function is a useful tool when small samples sizes are desired. When larger samples are necessary, we look to the Random Number Generation tool that Excel offers. It is explained below.

3.2.2 The Random Number Generation Tool

We will again utilize Example 3.22 found in the *Statistics for Business and Economics* text to demonstrate how to use the Random Number Generation tool within Excel. We begin by opening up Excel and placing the cursor on any cell in the blank worksheet. We click on the **Data** tab and then click on the **Data Analysis** Icon found in the **Analysis** group as shown below in Figure 3.5.

Figure 3.5

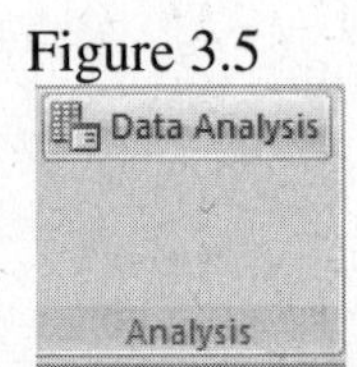

This will open up the **Data Analysis** menu shown below in Figure 3.6. Scroll down the list and select the **Random Number Generation** tool. Click OK.

Figure 3.6

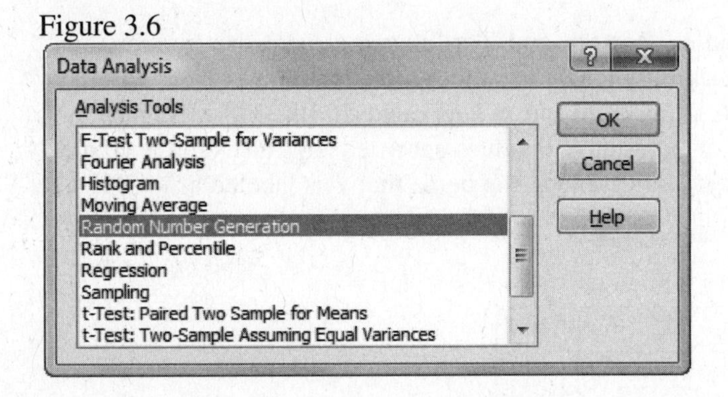

The Random Number Generation menu is opened as shown below in Figure 3.7. In this example, we are asked to take a single sample of five homes from the 100,000 homes available. We enter the value **1** as the **Number of Variables** to use, indicating we want just one sample. We enter the value **5** as the **Number of Random Numbers**, indicating our sample size. To choose a random sample as described in the text, we select the **Uniform Distribution** type and indicate the values **1** and **100,000** as the values to select **Between**. We have the option of choosing a specific random number sequence by entering a value in the **Random Seed** box, but we choose to let Excel decide and will leave it blank. The **Output Options** allow us to select where the random numbers will be located within the Excel workbook. In Figure 3.7 below, we have specified that Excel list the random numbers beginning in cell A1. We click on OK to select the random numbers

Figure 3.7

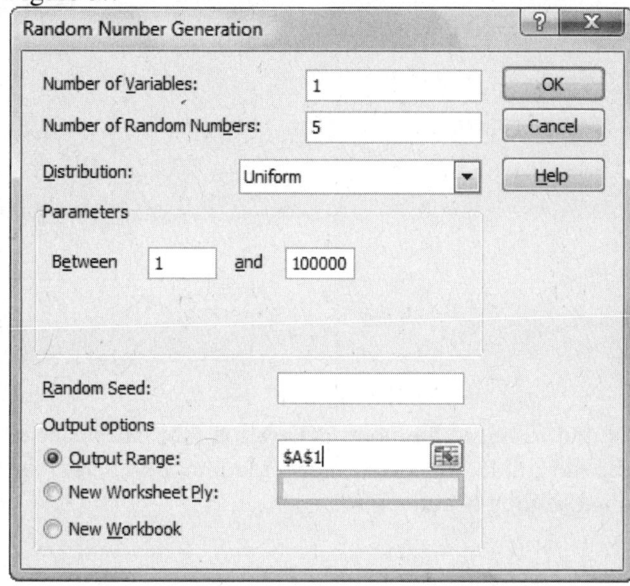

The random numbers selected by Excel are shown below in Figure 3.8. Notice that the random digits are not expressed as integers but contain decimal values as part of the random numbers.

Figure 3.8

A1

	A	B	C
1	1807.678		
2	68679.17		
3	78829.52		
4	67424.87		
5	40535.28		

To remove these decimals, we can format the numbers selected and decrease the decimals of the values until we are left with the rounded integers. Click on the **Home** tab and Select the **Decrease Integers** button found in the **Numbers** group as shown if Figure 3.9 below.

Figure 3.9

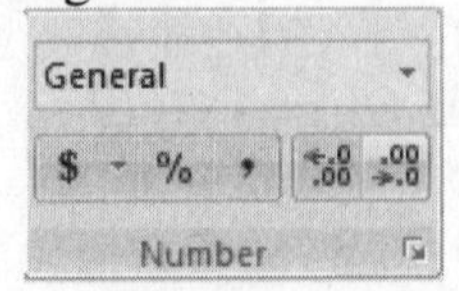

Continue clicking on the **Decrease Integers** button until all decimals are removed from the random numbers and you are left with values similar to those shown below in Figure 3.10. These would be the homes that should be sampled to generate a random sample of five homes from the 100,000 contained in the population.

Figure 3.10

A1

	A	B	C
1	1808		
2	68679		
3	78830		
4	67425		
5	40535		

It needs to be noted that both the RANDBETWEEN and Random Number Generation tool can result in repeated selections of random numbers. It may be desirable to select more random numbers than necessary to replace repeated random numbers when and if they occur.

Chapter 4
Random Variables and Probability Distributions

4.1 Introduction

Chapter 4 introduces random variables and sampling distributions to the reader. Three discrete random variables, the binomial, poisson, and hypergeometric distribution, are introduced. Three continuous random variables, the normal, uniform, and exponential distribution, are also introduced. In addition, assessing normality and the sampling distribution of the sample mean are introduced in Chapter 4.

Both individual and cumulative probabilities can be found for the discrete random variables using several different Excel functions. Individual probabilities are the exact probability of an outcome and will answer the "equal to" probability questions found in the text. Cumulative probabilities are the "at most" probabilities that are found in the cumulative tables listed in the text. Both are useful probabilities and will be explained below.

Excel functions can also be used in place of the statistical tables to find probabilities for the continuous normal and exponential distributions. The Excel normal distribution functions can be used both when working with the normal random variables found in Section 4.6 and also when working with the sampling distribution of the sample mean found in Section 4.11 of the text.

DDXL offers the user the ability to assess the normality of a distribution of data through the use of its normal probability plot option and through the various charting techniques already discussed in Chapter 2. We will describe the normal probability plot here in Chapter 4.

Finally, the uniform random variable can be solved rather easily using simple mathematical techniques. Neither Excel nor DDXL offers any functions or techniques to calculate these probabilities.

The following examples from *Statistics for Business and Economics* are solved with Microsoft Excel® in this chapter:

Excel Companion		**Statistics for Business and**	
Exercise	**Page**	**Economics**	**Excel File Name**
4.1	65	Example 4.12	
4.2	68	Example 4.13	
4.3	72	Example 4.14	
4.4	73	Example 4.20	
4.5	75	Example 4.23	
4.6	77	Example 4.27	
4.7	78	Example 4.24	EPAgas
4.8	81	Example 4.32	

4.2 Calculating Binomial Probabilities

To use the binomial probability tool within Excel, we begin by opening up Excel and placing the cursor on any cell in the blank worksheet. We click on the **Home** tab and then select the arrow next to the **Function** Icon in the **Editing** group of options shown below in Figure 4.1.

Figure 4.1

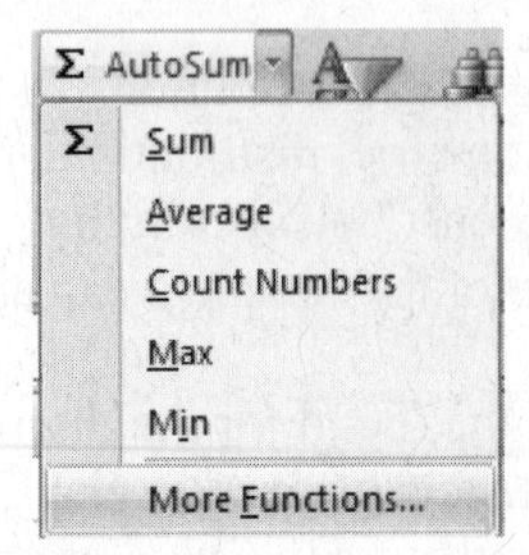

We then select the **More Functions** option to get the **Insert Function** menu shown below in Figure 4.2. Click on the arrow to select that **All** categories are being used and scroll down until you reach the **BINOMDIST** function. Click OK to open the **BINOMDIST** function.

Figure 4.2

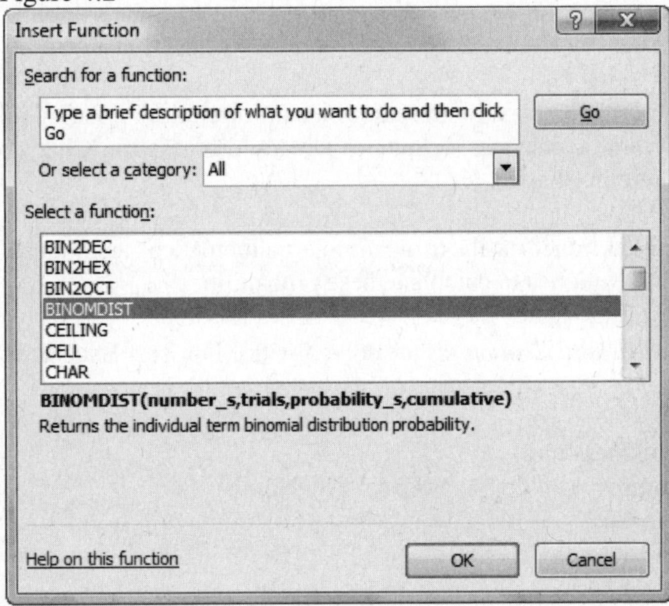

Figure 4.3

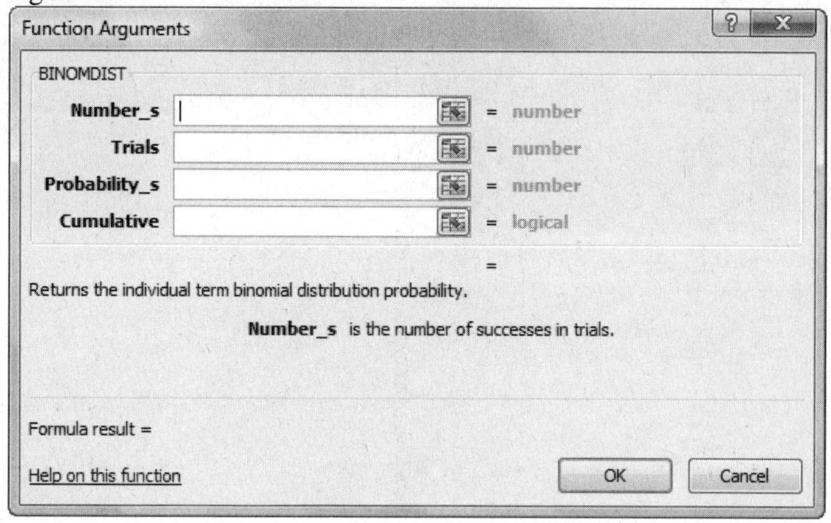

The **BINOMDIST** function requires the user to enter the number of successes (**Number_s**), the sample size (**Trials**), the probability of a success (**Probability_s**), and the type of probability desired (**Cumulative – either True or False**). For most applications, the cumulative probability option should be selected (**Cumulative** = **True**) in order to maximize the information that Excel will offer. **Click OK** to finish. We illustrate with the next example.

Exercise 4.1: As an example, we turn to Example 4.12 from the *Statistics for Business and Economics* text.

Suppose a poll of 20 employees is taken in a large company. The purpose is to determine x, the number who favor unionization. Suppose that 60% of all the company's employees favor unionization.

a. Find the mean and standard deviation of x.
b. Find the probability that $x \leq 10$.
c. Find the probability that $x > 12$.
d. Find the probability that $x = 11$.

Solution:

We utilize the **BINOMDIST** function to solve parts b-d of this problem. We identify in the problem that the sample size is **n=20** and the probability of a success is **p=.60**. We enter both of these values in the appropriate locations in the **BINOMDIST** menu shown if Figure 4.4. In order to solve part b, we need to find the cumulative probability for the value of x=10 successes. We specify the number of successes to be **x=10** and we enter **True** option in the Cumulative box to signify that we want a cumulative probability.

Figure 4.4

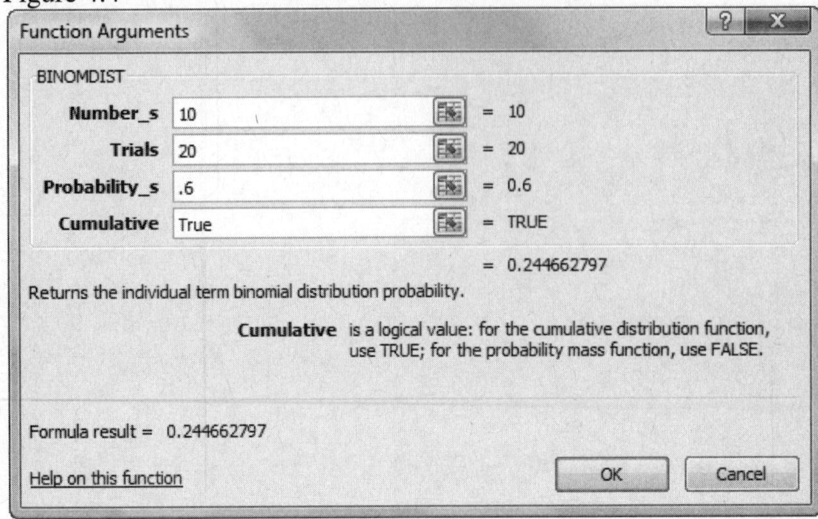

We note that the probability is shown in Figure 4.4 and then placed in our worksheet cell once we click OK. That probability is given as 0.244663.

Part c asks us to find the P(X > 12). As we saw in the text, there are two methods for finding this probability. We could use the **BINOMDIST** function and find the exact probabilities (Cumulative=False) for the values of x=13, 14, 15,…,20 and then add the results together. A smarter and more efficient method is to recognize that the probability that we want, P(X > 12), can be found by finding 1 – P(X ≤ 12). By using the cumulative feature of the **BINOMDIST**, we substitute the value of X = 12 into the Number_s box and find the result shown below.

Figure 4.5

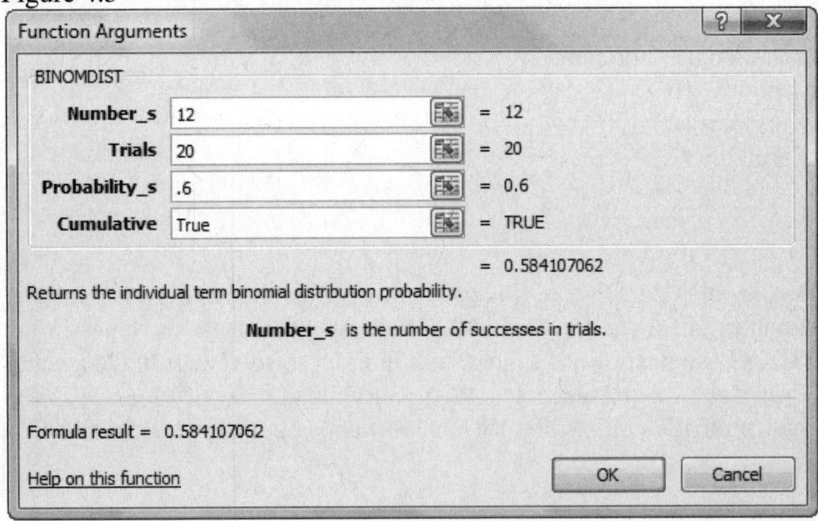

We find that $P(X > 12) = 1 - P(X \leq 12) = 1 - 0.58411 = 0.41589$.

Lastly, we find the exact probability asked for in part d by using the individual probability option (Cumulative=False). To find P(X = 11), we use 11 as the number of successes and select False for the Cumulative option. The results are shown in Figure 4.6.

Figure 4.6

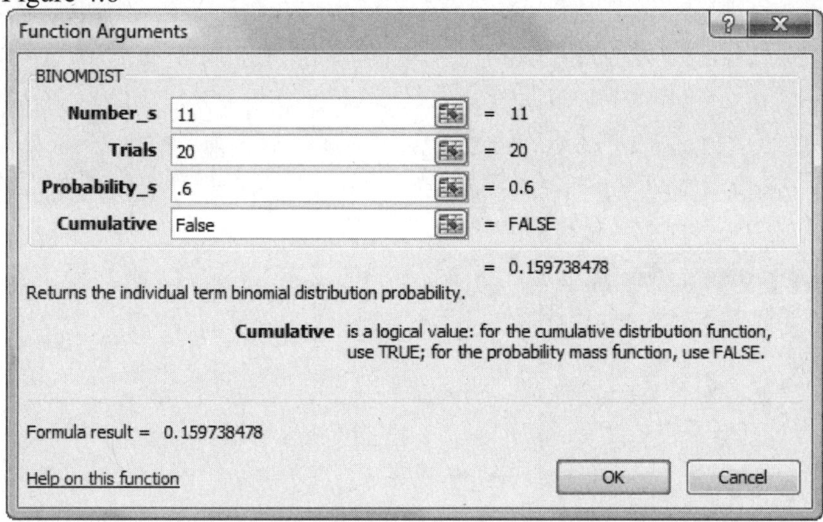

We find P(X = 11) = 0.15974.

4.3 Calculating Poisson Probabilities

To use the poisson probability tool within Excel, we begin by opening up Excel and placing the cursor on any cell in the blank worksheet. We click on the **Home** tab and then select the arrow next to the **Function** Icon in the **Editing** group of options shown below in Figure 4.7.

Figure 4.7

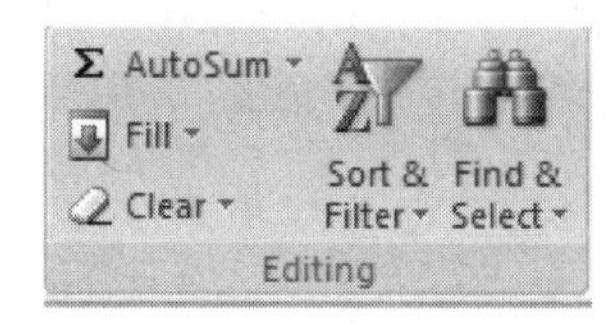

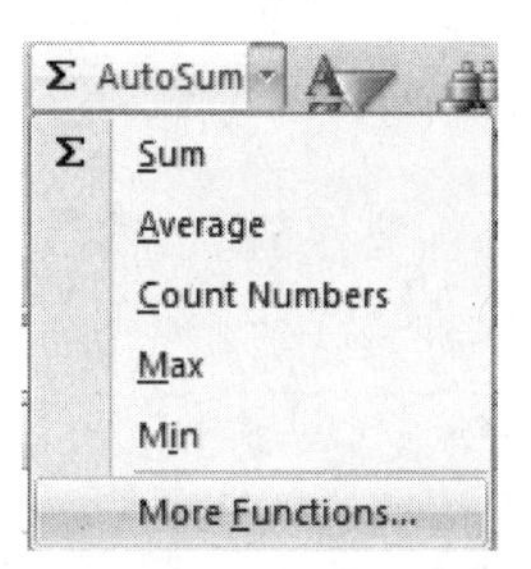

We then select the **More Functions** option to get the **Insert Function** menu shown below in Figure 4.8. Click on the arrow to select that **All** categories are being used and scroll down until you reach the **POISSON** function. Click OK to open the **POISSON** function.

Figure 4.8

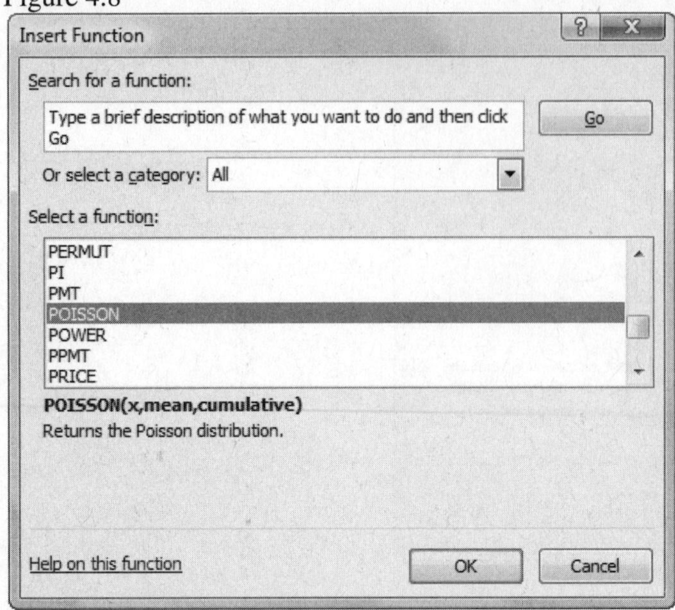

Figure 4.9

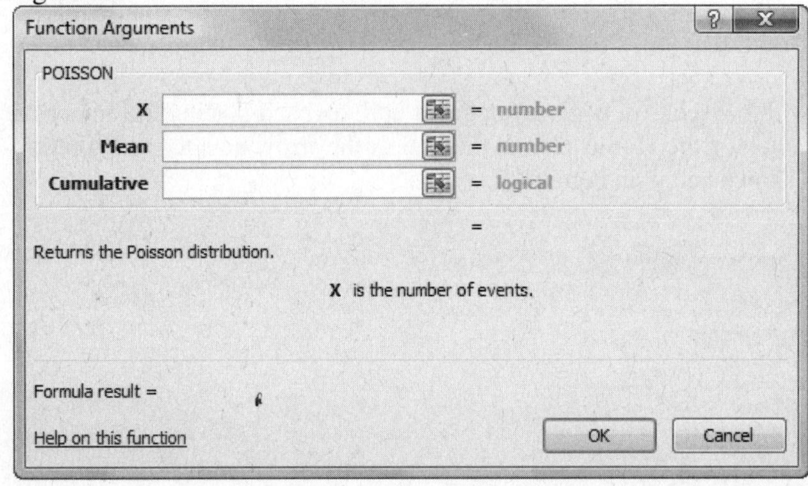

The **POISSON** function requires the user to enter a value of X to find a probability for, (**X**), the mean of the distribution (**Mean**)**,** and the type of probability desired (**Cumulative – either True or False**). For most applications, the cumulative probability option should be selected (**Cumulative = True**) in order to maximize the information that Excel will offer. **Click OK** to finish. We illustrate with the next example.

Exercise 4.2: As an example, we turn to Example 4.13 from the *Statistics for Business and Economics* text.

Suppose the number, x, of a company's employees who are absent on Mondays has (approx.) a Poisson probability distribution. Furthermore, assume that the average number of Monday absentees is 2.6.

a. Find the mean and standard deviation of x, the number of employees absent on Monday.
b. Find the probability that fewer than two employees are absent on a given Monday.
c. Find the probability that more than five employees are absent on a given Monday.
d. Find the probability that exactly five employees are absent on a given Monday.

Solution:

We utilize the **POISSON** function to solve parts b-d of this problem. We identify in the problem that the mean of the distribution is Mean=2.6. In order to solve part b, we need to find the cumulative probability for the value of x=1(since $P(X < 2) = P(X \leq 1)$ and we can utilize the cumulative function). We specify the **x=1** and we enter **True** option in the Cumulative box to signify that we want a cumulative probability.

Figure 4.10

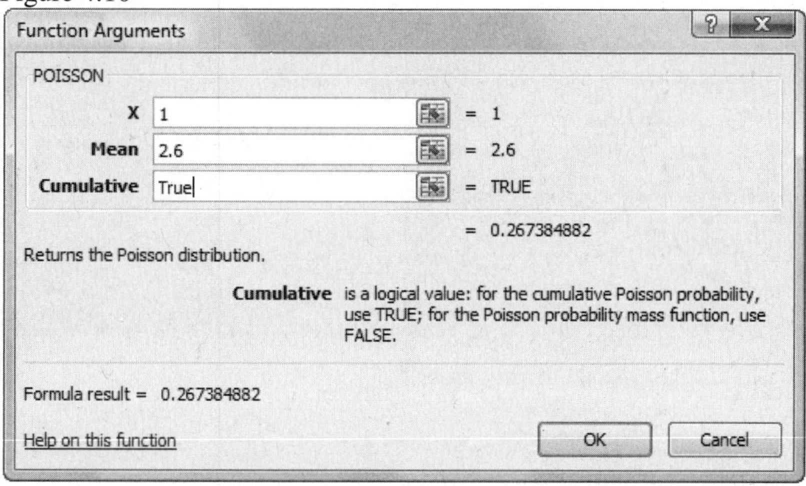

We note that the probability is shown in Figure 4.10 and then placed in our worksheet cell once we click OK. That probability is given as 0.267395.

Part c asks us to find the $P(X > 5)$. As we saw in the text, there are two methods for finding this probability. We could use the **POISSON** function and find the exact probabilities (Cumulative=False) for the values of x=5, 6, 7, 8, … and then add the results together. A smarter and more efficient method is to recognize that the probability that we want, $P(X > 5)$, can be found by finding $1 - P(X \leq 5)$. By using the cumulative feature of the **POISSON** function, we substitute the value of **X=5** into the proper location and find the result shown below.

Figure 4.11

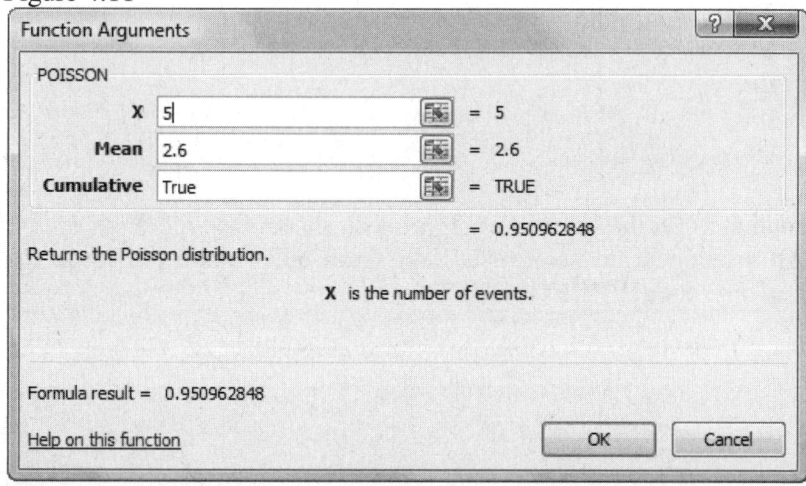

We find that $P(X > 5) = 1 - P(X \leq 5) = 1 - 0.950963 = 0.049037$.

Lastly, we find the exact probability asked for in part d by using the individual probability option (Cumulative=False). To find P(X = 5), we use 5 as the number of successes and select False for the Cumulative option. The results are shown in Figure 4.12.

Figure 4.12

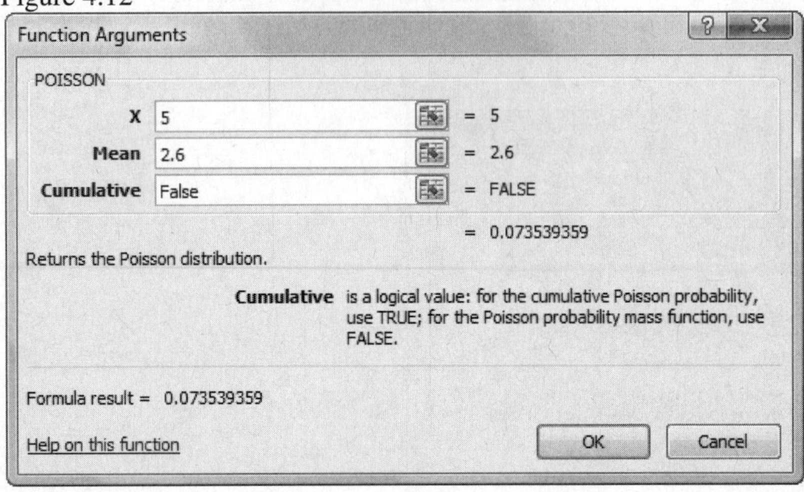

We find P(X = 5) = 0.073539.

4.4 Calculating Hypergeometric Probabilities

To use the hypergeometric probability tool within Excel, we begin by opening up Excel and placing the cursor on any cell in the blank worksheet. We click on the **Home** tab and then select the arrow next to the **Function** Icon in the **Editing** group of options shown below in Figure 4.13.

Figure 4.13

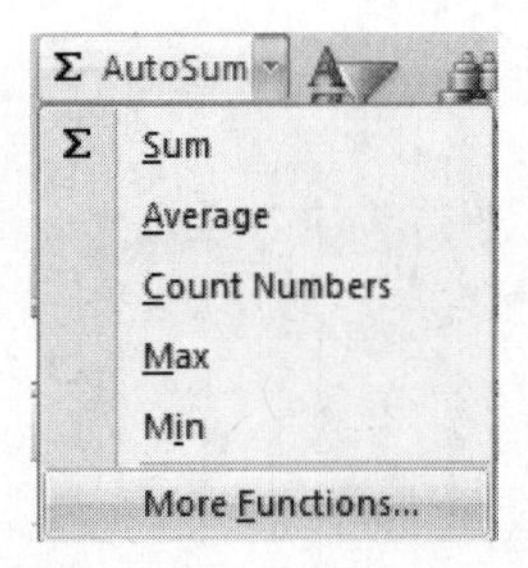

We then select the **More Functions** option to get the **Insert Function** menu shown below in Figure 4.14. Click on the arrow to select that **All** categories are being used and scroll down until you reach the **Hypergeometric** function. Click OK to open the **HYPGEOMDIST** function.

Figure 4.14

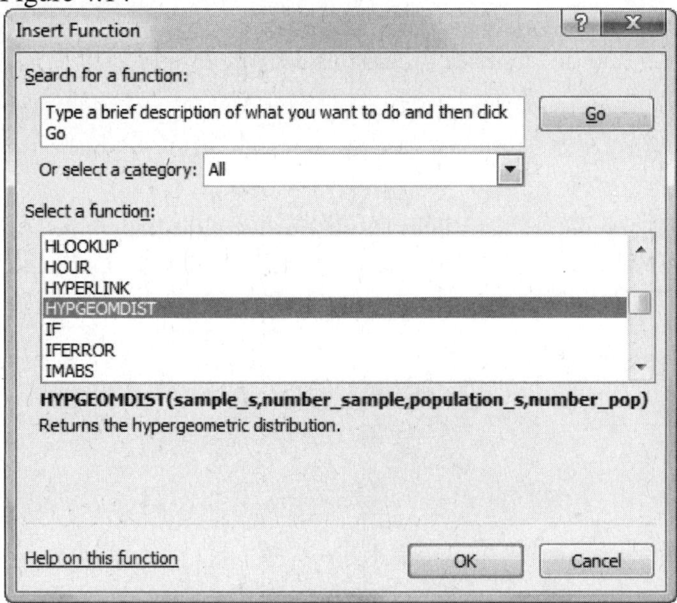

Figure 4.15

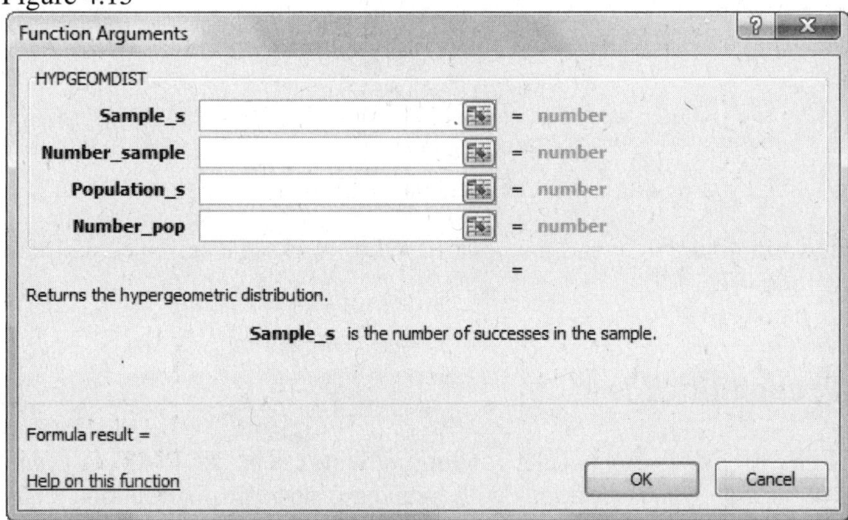

The **HYPGEOMDIST** function requires the user to enter the size of the population (**Number_pop**), the number of successes in the population (**Population_s**), the sample size (**Number_sample**), and the number of successes in the sample (**Sample_s**). We illustrate with the next example.

Exercise 4.3: As an example, we turn to Example 4.14 from the *Statistics for Business and Economics* text.

Suppose a marketing professor randomly selects three new teaching assistants from a total of 10 applicants – six male and four female students. Let x be the number of females who are hired.

a. Find the mean and standard deviation of x.
b. Find the probability that no females are hired.

To find the probability in part b, we use the hypergeometric distribution with $N = 10$ (**Number_pop=10**), $n = 3$ (**Number_sample=3**), and $r = 4$ (**Population_s=4**). To find the probability that no females are hired, we use $x = 0$ (**Sample_s=0**) in the **HYPGEOMDIST** menu shown in Figure 4.16.

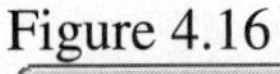
Figure 4.16

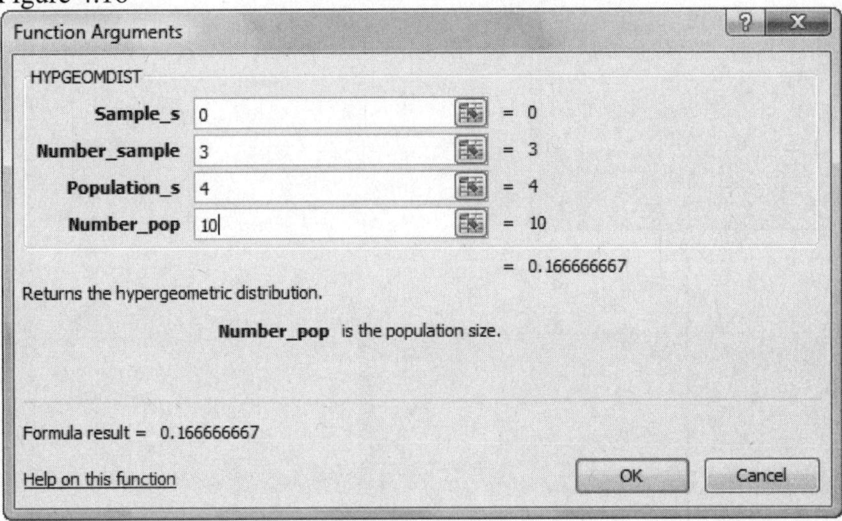

We note that the probability is shown in Figure 4.16 and then placed in our worksheet cell once we click OK. That probability is given as 0.1666667.

4.5 Calculating Normal Probabilities

Excel offers four different functions to work with normal random variables. The NORMSDIST and NORMSINV functions are only useful when working with standard normal distributions. The **NORMDIST** and **NORMINV** functions are used when working with any normal distributions. Because most normal applications utilize non-standard normal distributions, we will illustrate how to work with the **NORMDIST** and **NORMINV** functions. If a standard normal distribution is needed, the user can enter the values of the mean = 0 and the standard deviation = 1 and use these functions to generate the required information. As an alternative, the user could utilize the NORMSDIST and NORMSINV functions when working with standard normal distributions.

To use the normal probability tools within Excel, we begin by opening up Excel and placing the cursor on any cell in the blank worksheet. We click on the **Home** tab and then select the arrow next to the **Function** Icon in the **Editing** group of options shown below in Figure 4.17.

Figure 4.17

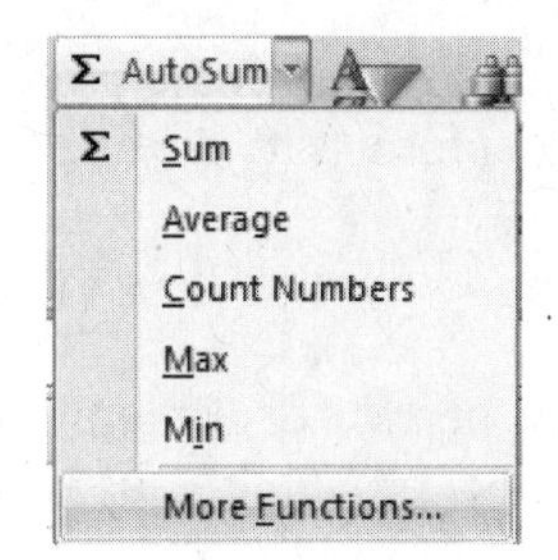

We then select the **More Functions** option to get the **Insert Function** menu shown below in Figure 4.18. Click on the arrow to select that **All** categories are being used and scroll down until you reach the **NORMDIST** and **NORMINV** functions. Click OK to open the function needed.

Figure 4.18

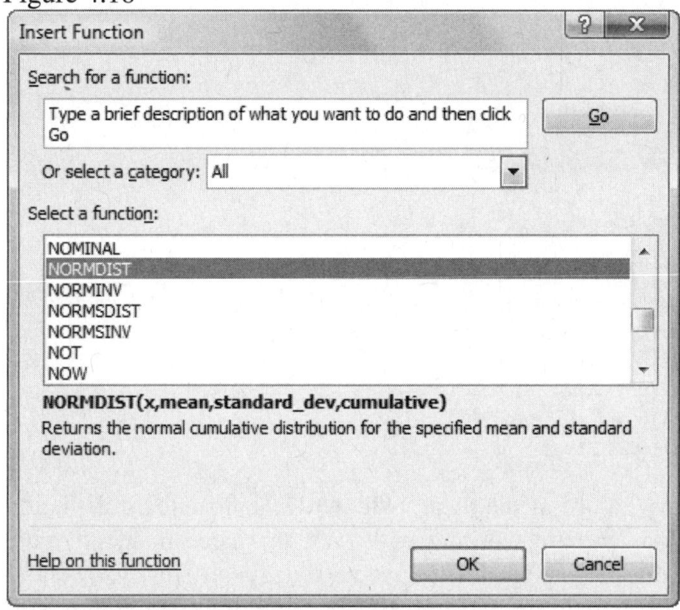

The **NORMDIST** function allows the user to determine the cumulative probability of a point in the normal distribution, given values of the mean and standard deviation of that distribution. The **NORMINV** function allows the user to determine the point in the normal distribution at which a specified cumulative probability is achieved, given values of the mean and standard deviation of that distribution. It is up to the user to determine which function is appropriate based on the type of information is desired. We illustrate both procedures below using two separate examples taken from the text.

Exercise 4.4: We use Example 4.20 from the *Statistics for Business and Economics* text:

Suppose an automobile manufacturer introduces a new model that has an advertised mean in-city mileage of 27 miles per gallon. Although such advertisements seldom report in any measure of variability, suppose you write the manufacturer for the details of the test, and find that the standard deviation is 3 miles per gallon. This information leads you to formulate a probability model for the random variable, x, the in-city mileage for this car model. You believe that the probability distribution of x can be approximated by a normal distribution with a mean of 27 and a standard deviation of 3.

a. If you were to buy this model of automobile, what is the probability that you would purchase one that averages less than 20 miles per gallon for in-city driving? In other words, find $P(x < 20)$.

Solution:

We first identify that we are working with the normal distribution with a mean of 27 and a standard deviation of 3. We are asked to determine the probability of selecting a value from that normal distribution that falls below the value 20. The **NORMDIST** function is appropriate to use for this problem. We select the **NORMDIST** function from our list of functions and click **OK**.

Figure 4.19

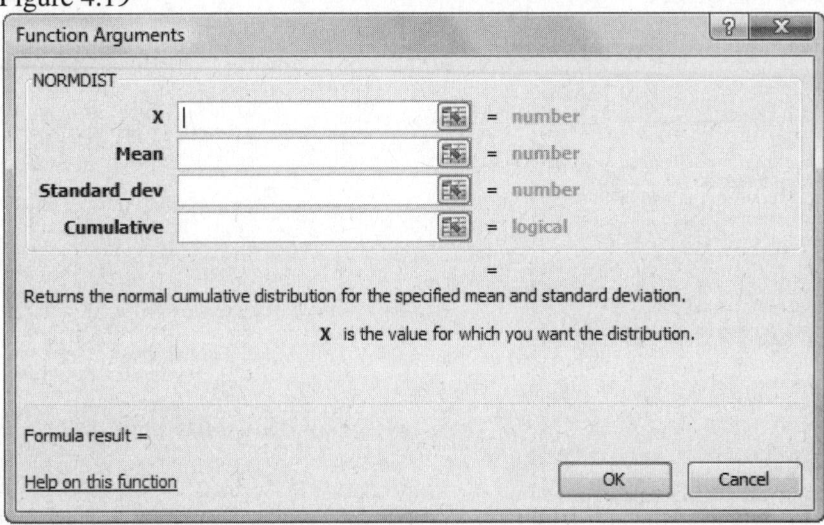

The **NORMDIST** function requires us to identify values of the mean (**Mean=27**) and standard deviation (**Standard_dev=3**) of the normal distribution that we are working with. We also need to identify the value (**X=20**) in the distribution that we want probabilities for. Lastly, we need to specify that we want to work with cumulative probabilities by entering **TRUE** in the **Cumulative** box in the menu. Figure 4.20 shows the completed function.

Figure 4.20

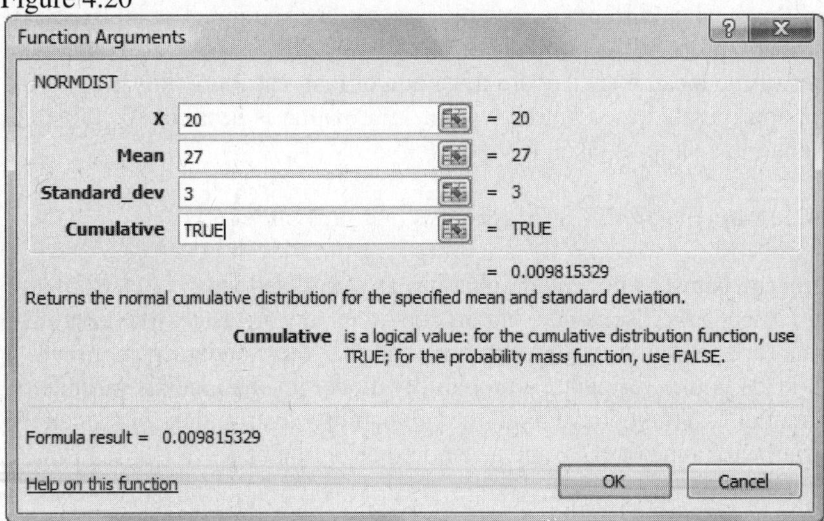

We note that the probability is shown in Figure 4.20 and then placed in our worksheet cell once we click OK. That probability is given as 0.00982.
It is important to remember that the **NORMDIST** function will only report cumulative probabilities. While we desired to find P(X < 20) in this problem, many problems ask us to find different types of probabilities. The user needs to be able to use the cumulative probabilities given by Excel to find other probabilities that are desired.

For example, if we wanted to determine P(X > 20), we would take the cumulative probability found in Excel, and subtract it from the value one.

$P(X > 20) = 1 - P(X \leq 20) = 1 - 0.00982 = .99018.$

The **NORMDIST** function can be used in this manner to determine any probabilities desired from normal distributions. To find specific values of a normal distribution that occur at a given probability, we use the **NORMINV** function described in the following exercise.

Exercise 4.5: We use the Example 4.23 from the *Statistics for Business and Economics* text:

Suppose a paint manufacturer has a daily production, x, that is normally distributed with a mean of 100,000 gallons and a standard deviation of 10,000 gallons. Management wants to create an incentive bonus for the production crew when the daily production exceeds the 90th percentile of the distribution, in hopes that the crew will, in turn, become more productive. At what level of production should management pay the incentive bonus?

Solution:

We first identify that we are working with the normal distribution with a mean of 100,000 and a standard deviation of 10,000. We are asked to determine the value in the distribution that exceeds the 90th percentile. The **NORMINV** function is appropriate to use for this problem. We select the **NORMINV** function from our list of functions and click **OK**.

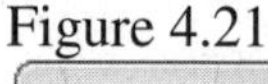
Figure 4.21

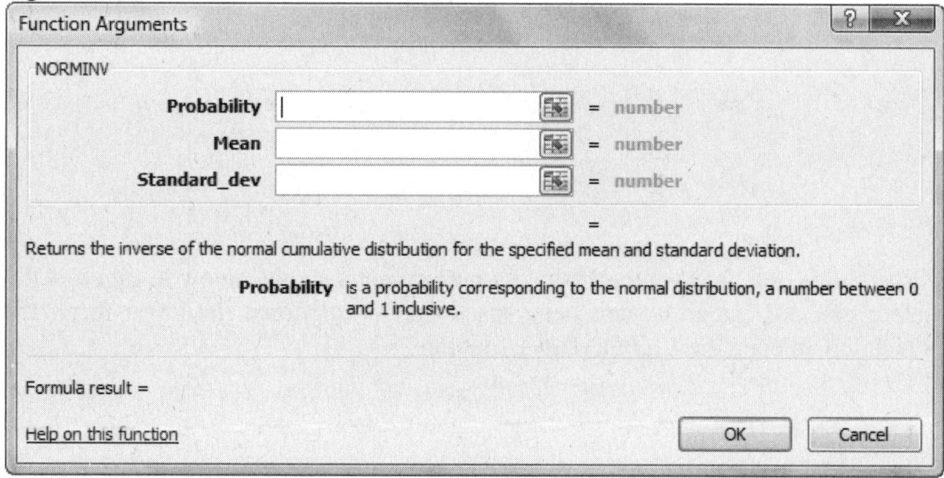

The **NORMINV** function requires us to identify values of the mean (**Mean=27**) and standard deviation (**Standard_dev=3**) of the normal distribution that we are working with. We also need to identify the cumulative probability that applies to the point in the distribution that we are trying to find (**Probability=.90**). Figure 4.22 shows the completed function.

Figure 4.22

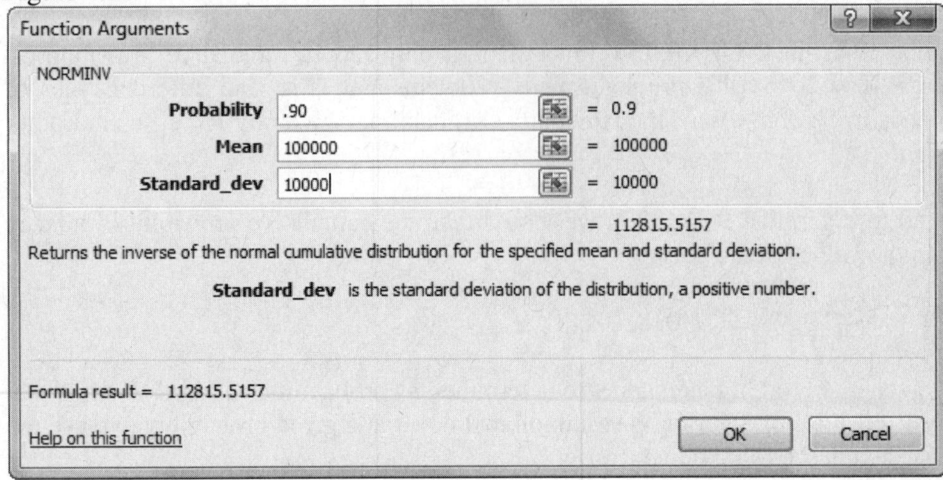

We note that the desired value is shown in Figure 4.22 and then placed in our worksheet cell once we click OK. That value is given as 112,815.5157.

It is important to note that the **NORMINV** function only works with cumulative probabilities. When problems express other types of probabilities, the user must use the given probabilities to find the cumulative probabilities required in the **NORMINV** function.

4.6 Calculating Exponential Probabilities

To use the exponential probability tool within Excel, we begin by opening up Excel and placing the cursor on any cell in the blank worksheet. We click on the **Home** tab and then select the arrow next to the **Function** Icon in the **Editing** group of options shown below in Figure 4.23.

Figure 4.23

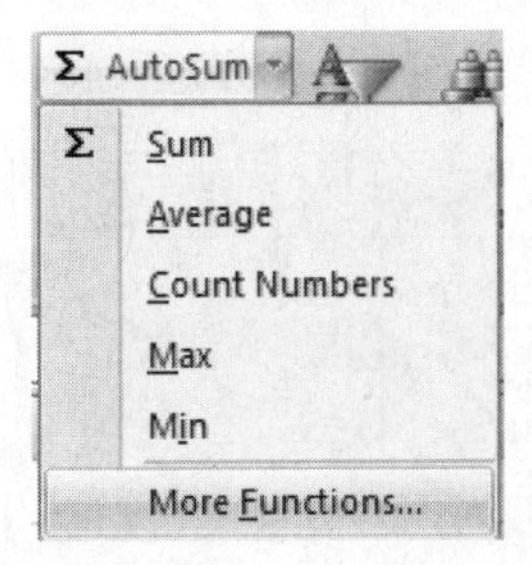

We then select the **More Functions** option to get the **Insert Function** menu shown below in Figure 4.24. Click on the arrow to select that **All** categories are being used and scroll down until you reach the **EXPONDIST** function. Click OK to open the **EXPONDIST** function.

Figure 4.24

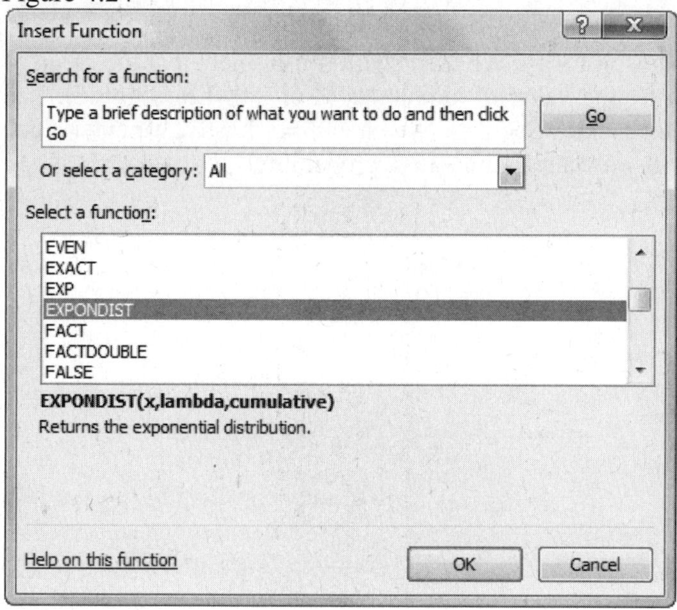

Figure 4.25

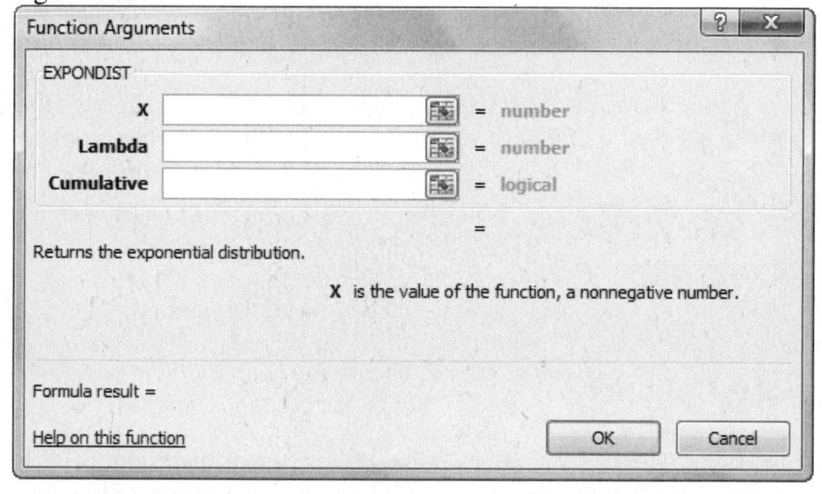

The **EXPONDIST** function requires the user to enter a value of X to find a probability for, (**X**), the value of lambda (where $lambda = \frac{1}{mean}$) for the distribution (**Lamda),** and the type of probability desired (**Cumulative – either True or False**). For most applications, the cumulative probability option should be selected (**Cumulative = True**) in order to maximize the information that Excel will offer. **Click OK** to finish. We illustrate with the next example.

Exercise 4.6: As an example, we turn to Example 4.27 from the *Statistics for Business and Economics* text.

Suppose the length of time (in hours) between emergency arrivals at a certain hospital is models as an exponential distribution with a mean of 2. What is the probability that more than 5 hours pass without an emergency arrival?

Solution:

We utilize the **EXPONDIST** function to solve this problem. We identify in the problem that the mean of the distribution is the value 2. Therefore we enter the value of lambda as ½ or .50 (**Lambda=.5)**. The value of X we want to determine a probability for is the value 5, so we enter **X=5**. Lastly, we enter **True** as the option in the Cumulative box to signify that we want a cumulative probability.

Figure 4.26

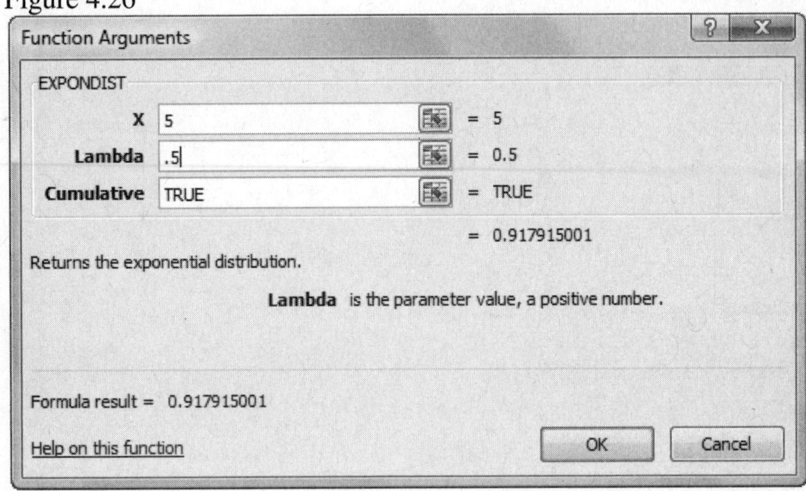

We note that the probability is shown in Figure 4.26 and then placed in our worksheet cell once we click OK. That probability is given as 0.917915. We note that this is a cumulative probability. In the problem, we wanted to find out the probability that more than 5 hours pass without an emergency. We use the computed probability to find:

$$P(X > 5) = 1 - P(X \leq 5) = 1 - .917915 = .082085$$

4.7 Assessing the Normality of a Data Set

DDXL offers the user a method of assessing whether a data set possesses a normal distribution. The Normal Probability Plot utility creates a plot that enables the reader to determine the shape of the data. **The DDXL Add-In** offers an easy method of creating this plot within the **Charts and Plots** menu. We will utilize the **Normal Probability Plot** option within this menu to solve the following problem.

Exercise 4.7: We utilize the EPA Gas Mileage Ratings for 100 Cars that is given in Example 4.24 of the *Statistics for Business and Economics* text. The data is shown below in Table 4.1. Construct a normal probability plot of the data and assess the shape of the EPA gas mileage ratings.

Table 4.1

EPA Gas Mileage Ratings for 100 Cars (miles per gallon)									
36.3	41.0	36.9	37.1	44.9	36.8	30.0	37.2	42.1	36.7
32.7	37.3	41.2	36.6	32.9	36.5	33.2	37.4	37.5	33.6
40.5	36.5	37.6	33.9	40.2	36.4	37.7	37.7	40.0	34.2
36.2	37.9	36.0	37.9	35.9	38.2	38.3	35.7	35.6	35.1
38.5	39.0	35.5	34.8	38.6	39.4	35.3	34.4	38.8	39.7
36.3	36.8	32.5	36.4	40.5	36.6	36.1	38.2	38.4	39.3
41.0	31.8	37.3	33.1	37.0	37.6	37.0	38.7	39.0	35.8
37.0	37.2	40.7	37.4	37.1	37.8	35.9	35.6	36.7	34.5
37.1	40.3	36.7	37.0	33.9	40.1	38.0	35.2	34.8	39.5
39.9	36.9	32.9	33.8	39.8	34.0	36.8	35.0	38.1	36.9

Solution:

We need to utilize the normal probability plot utility within the **DDXL** program. Before we begin, we must access the data set for this example. **Open** the Data File **EPAgas** by following the directions found in the preface of this manual. If done correctly, the data should appear in a workbook similar to that shown below in Figure 4.27. Use the mouse to select the data shown in the workbook.

Figure 4.27

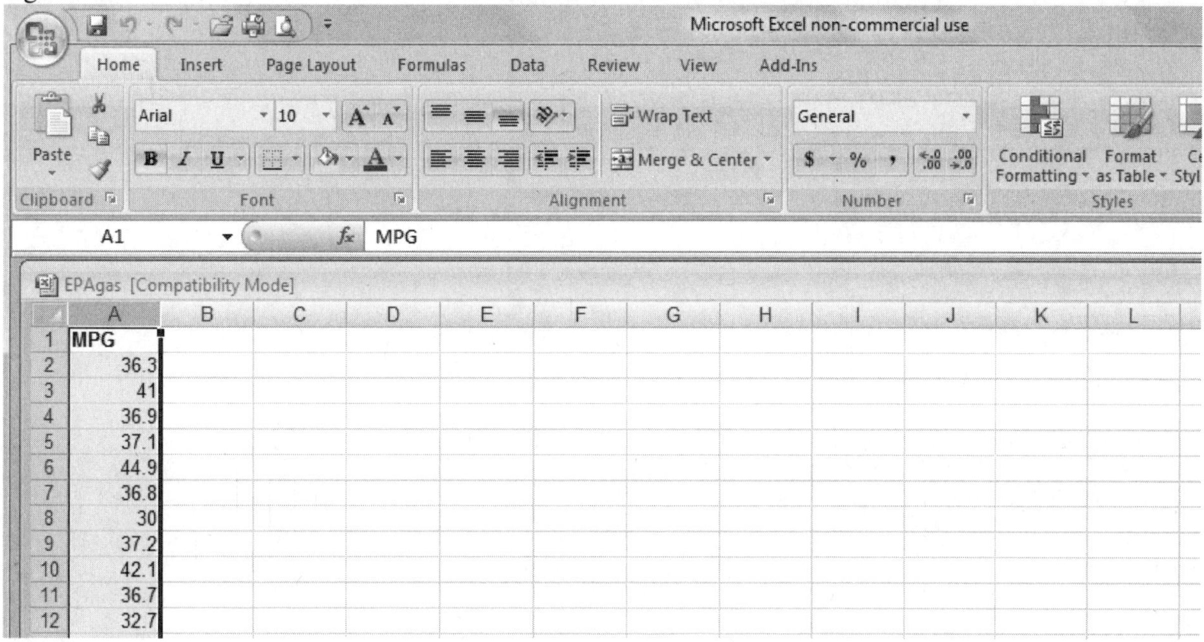

Click on the **DDXL Add-In** menu. Click on the **Charts and Plots** option to access the **Charts and Plots Dialog** menu (see Figure 4.28 below). Click on the ▾ in the Function Type box to access the different charts available to select. Highlight the **Normal Probability Plot** option to access the Normal Probability Plot menu shown in Figure 4.29. Highlight the **MPG** variable found in the **Names and Columns Box**.

Click on the ◀ to the right of the **Quantitative Variable** box to create a normal probability plot for the Lost variable. Click **OK**. The plot is shown in Figure 4.30.

Figure 4.28

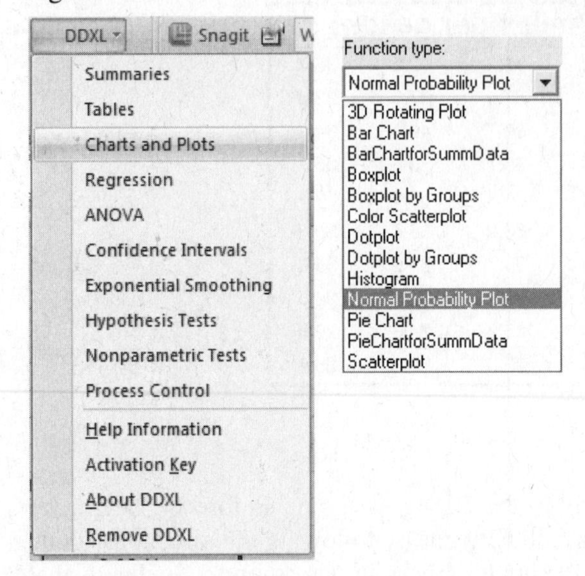

Figure 4.29

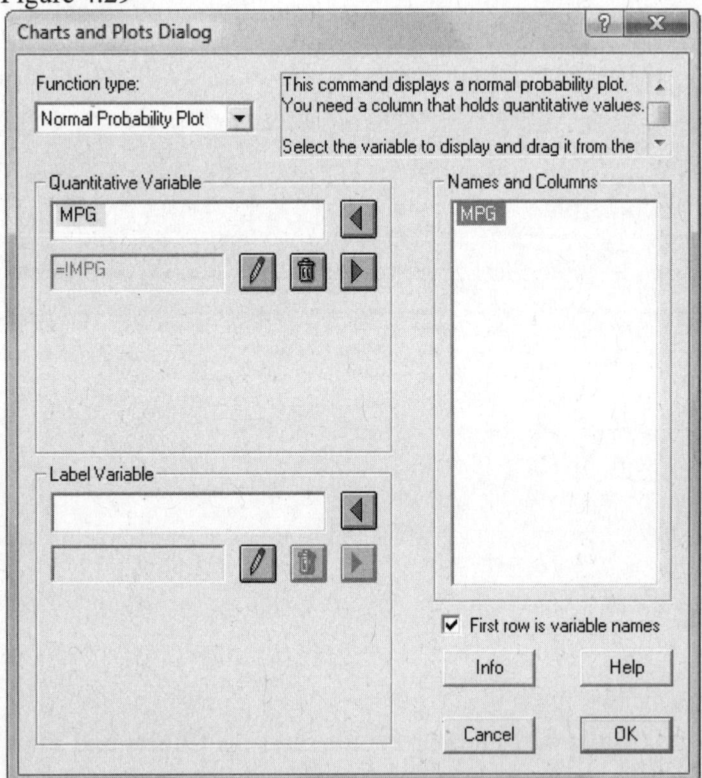

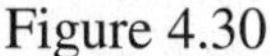
Figure 4.30

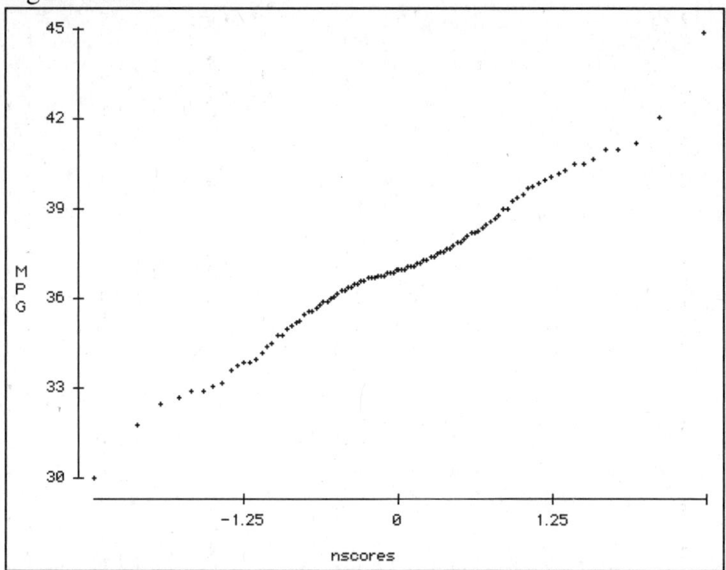

The straight line shown in the plot indicates that the data are extremely normal.

4.8 Calculating Probabilities Using the Sampling Distribution of $\bar{x}$

The Central Limit Theorem guarantees that for large n, the sampling distribution of the sample mean possesses an approximate normal sampling distribution. In order to calculate probabilities for these sampling distributions, we must utilize the normal probability distribution functions that we discussed in Section 4.5. We illustrate with the following example.

Exercise 4.8: We use Example 4.32 from the *Statistics for Business and Economics* text.

Suppose we have selected a random sample of n = 36 observations from a population with mean equal to 80 and standard deviation equal to 6. It is known that the population is not extremely skewed. Find the probability that the sample mean will be larger than 82.

Solution:

We first identify that we are working with the normal distribution with a mean of $\mu_{\bar{x}} = 80$ and a standard deviation of $\sigma_{\bar{x}} = \sigma/\sqrt{n} = 6/\sqrt{36} = 1$. We are asked to determine the probability of selecting a value from this normal distribution that falls larger than the value 82. The **NORMDIST** function is appropriate to use for this problem. We select the **NORMDIST** function from our list of functions and click **OK**.

Figure 4.31

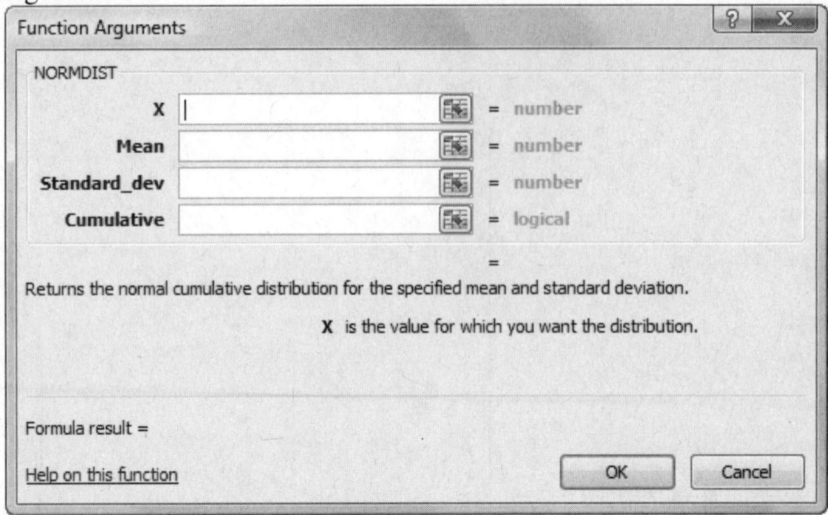

The **NORMDIST** function requires us to identify values of the mean (**Mean=80**) and standard deviation (**Standard_dev=1**) of the normal distribution that we are working with. We also need to identify the value (**X=82**) in the distribution that we want probabilities for. Lastly, we need to specify that we want to work with cumulative probabilities by entering **TRUE** in the **Cumulative** box in the menu. Figure 4.32 shows the completed function.

Figure 4.32

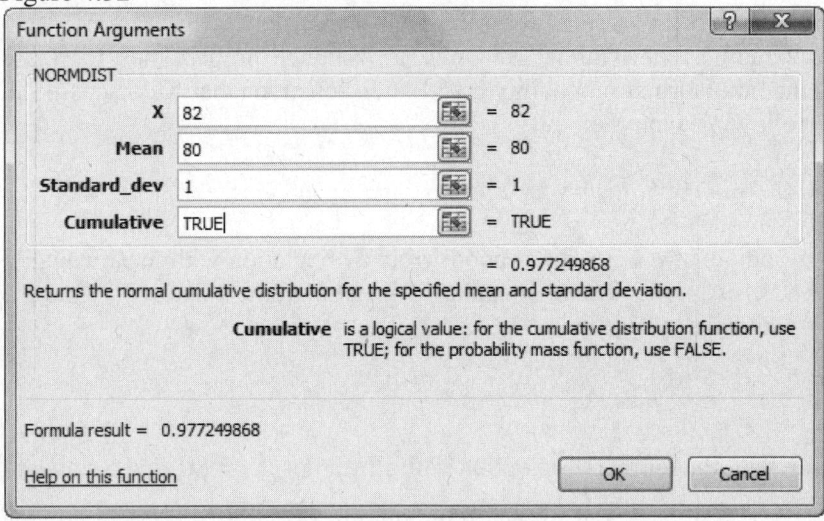

We note that the probability is shown in Figure 4.32 and then placed in our worksheet cell once we click OK. That probability is given as 0.97725. This cumulative probability can be used to find the probability we desire as follows:

$$P(X > 82) = 1 - P(X \leq 82) = 1 - 0.97725 = .02275$$

Technology Lab

The following exercises have been taken from *Statistics for Business and Economics* for you to practice the techniques discussed in this chapter. The output generated from these problems is also given for you to check your work.

4.67 U.S. airlines average about 4.5 fatalities per month (*Statistical Abstract of the United States: 2008*). Assume the probability distribution for x, the number of fatalities per month, can be approximated by a Poisson probability distribution.

a. What is the probability that no fatalities will occur during any given month?
b. What is the probability that one fatality will occur during a month?
c. Find $E(x)$ and the standard deviation of x.

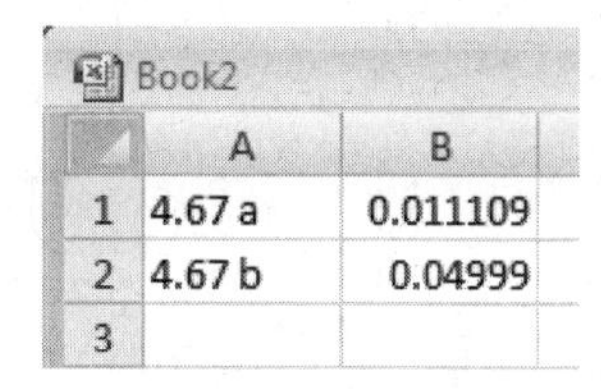
Book2

	A	B
1	4.67 a	0.011109
2	4.67 b	0.04999
3		

4.73 Suppose that you are purchasing cases of wine (12 bottles per case) and that, periodically, you select a test case to determine the adequacy of the bottles' seal. To do this, you randomly select and test 3 bottles in the case. If a case contains 1 spoiled bottle of wine, what is the probability that this bottle will turn up in your sample?

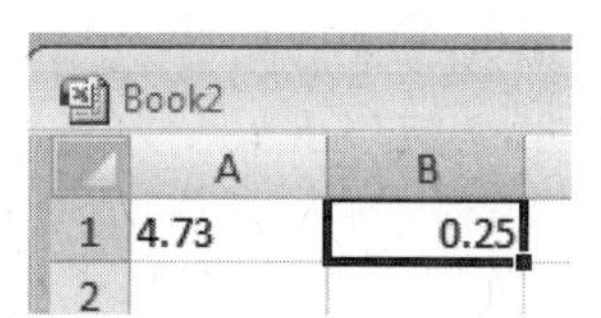
Book2

	A	B
1	4.73	0.25
2		

4.99 Personnel tests are designed to test a job applicant's cognitive and or physical abilities. The Wonderlic IQ test is an example of the former; a speed test involving the arrangement of pegs on the Purdue Pegboard is an example of the latter. A particular dexterity test is administered nationwide by a private testing service. It is known that for all tests administered last year the distribution of scores was approximately normal with mean 75 and standard deviation 7.5.

a. A particular employer requires job candidates to score at least 80 on the dexterity test. Approximately what percentage of the test scores during the past year exceeded 80?

b. The testing service reported to a particular employer that one of its job candidate's scores fell at the 98th percentile of the distribution (i.e., approximately 98% of the scores were lower than the candidate's, and only 2% were higher). What was the candidates score?

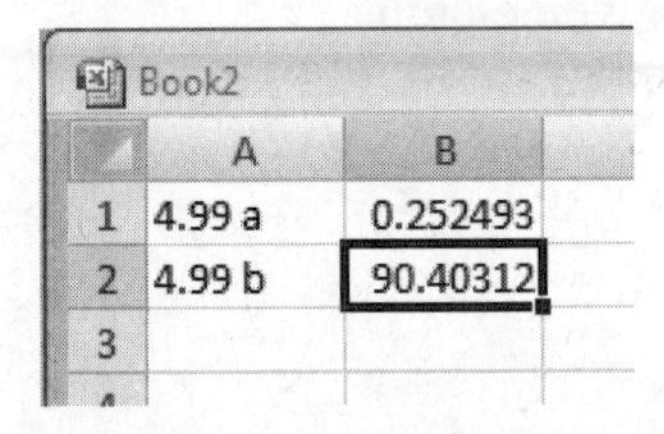
Book2

	A	B
1	4.99 a	0.252493
2	4.99 b	90.40312
3		

4.148 In the National Hockey League (NHL), games that are tied at the end of three periods are sent into "sudden-death" overtime. In overtime, the team to score the first goal wins. An analysis of NHL overtime games showed that the length of time elapsed before the winning goal is scored has an exponential distribution with mean 9.15 minutes (*Chance*, Winter 1995).

a. For a randomly selected overtime NHL game, find the probability that the winning goal is scored in 3 minutes or less.

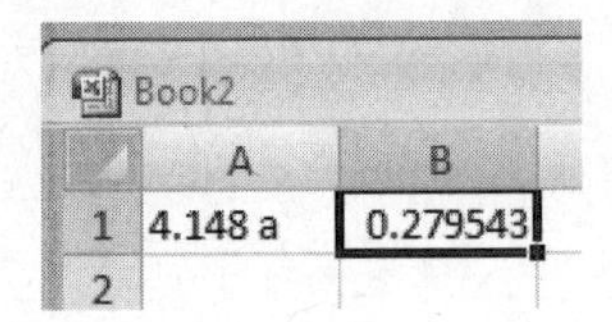
Book2

	A	B
1	4.148 a	0.279543
2		

4.172 According to a National Business Travel Association (NBTA) 2008 survey, the average salary of a travel management professional is \$97,300. Assume that the standard deviation of such salaries is \$30,000. Consider a sample of 50 travel management professionals and let $\overline{x}$ represent the mean salary for the sample.

d. Find $P(\overline{x} > 89{,}500)$

Book2

	A	B
1	4.172 e	0.967004
2		

4.219 *Whistle blowing* refers to an employee's reporting of wrongdoing by coworkers. The survey found that about 5% of employees contacted had reported wrongdoing in the past 12 months. Assume that a sample of 25 employees in one agency are contacted, and let x be the number who have observed and reported wrongdoing in the last 12 months. Assume that the probability of whistle blowing is .05 for any federal employee over the past 12 months.

a. Find the mean and standard deviation of x.
b. Find the probability that at least five of the employees are whistle blowers.

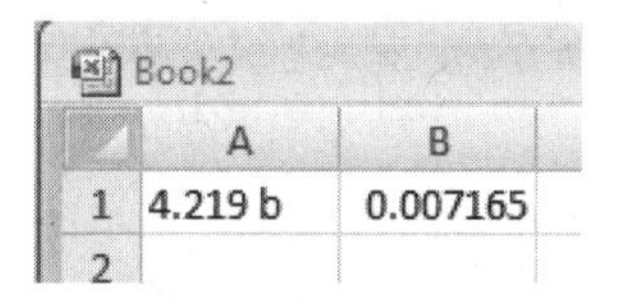
Book2

	A	B
1	4.219 b	0.007165
2		

Chapter 5
Inferences Based on a Single Sample: Confidence Intervals

5.1 Introduction

Chapter 5 introduces the reader to estimating population parameters with confidence intervals. Two parameters, the population mean and the population proportion, are studied in the chapter. The reader is also introduced to the topic of sample size determination, as it follows nicely from the estimation material presented.

DDXL provides calculation of confidence intervals for both means and proportions. There are two procedures presented for estimating a population mean; when the population standard deviation is known, and when the population standard deviation is unknown. Since the population standard deviation is almost never known, we concentrate our work on the unknown standard deviation case.

The confidence interval for a population proportion requires the user to specify the value of the qualitative variable that represents a success. DDXL then computes a confidence interval for the proportion of successes in the population.

DDXL restricts the user to specifying the reliability levels 90%, 95%, or 99% for both confidence intervals for means and proportions. Excel offers a data analysis tool that will allow the user to create confidence intervals for population means using any level of reliability they desire. With some creative data manipulation, the user can use this tool to create confidence intervals for population proportions as well.

The following examples from *Statistics for Business and Economics* illustrate the confidence interval calculations that can be found using DDXL and Excel in this chapter:

Excel Companion Exercise	Page	Statistics for Business and Economics Example	Excel File Name
5.1	88	Example 5.5	PRINTHEAD
5.2	91	Example 5.7	POLLING

5.2 Estimation of a Population Mean - Sigma Unknown

When estimating a population mean, it is highly unlikely that the population standard deviation will be known. In such cases, it is necessary to estimate the value of the population standard deviation using the sample standard deviation. This estimated standard deviation is then used to create the confidence interval for the population mean. DDXL uses this type of technique to create confidence intervals for population means.

To use the estimation tool within DDXL, **open** a new workbook and place the cursor in the upper left cell of the worksheet. Click on the **DDXL Add-In** menu. Click on the **Confidence Intervals** option to access the **Confidence Intervals** menu (see Figure 5.1 below). Click on the ▾ in the Function Type box to access the different confidence inteervals available to select. Highlight the **1 Var t Interval** option to access the Confidence Interval Dialog menu shown in Figure 5.2.

Figure 5.1

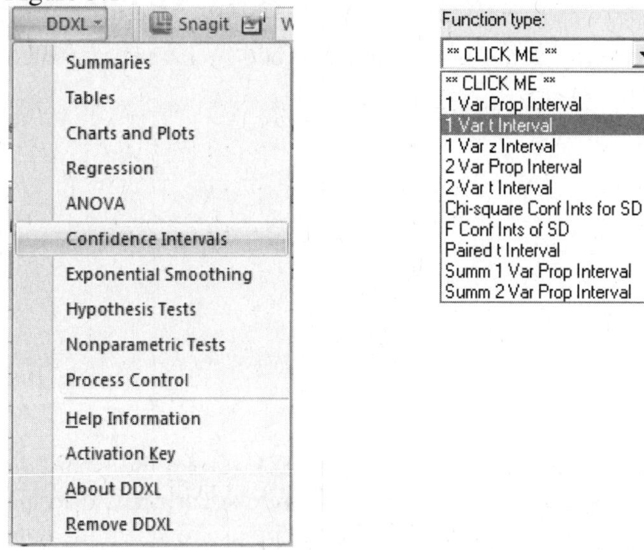

Figure 5.2

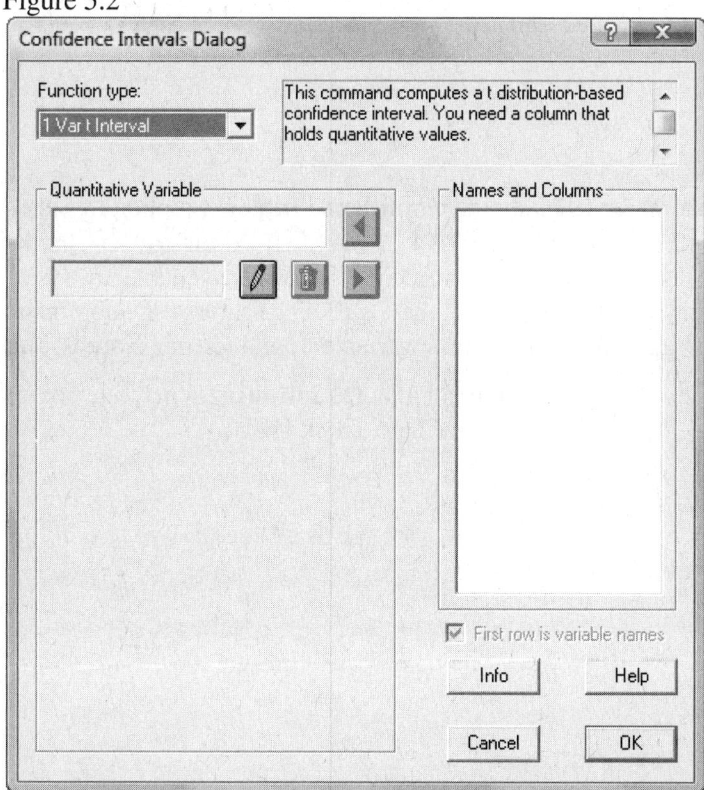

We illustrate how to use this technique with the following exercise:

Exercise 5.1: We use Example 5.5 found in the *Statistics for Business and Economics* text.

Some quality control experiments require destructive sampling (i.e., the test to determine whether the item is defective destroys the item) in order to measure some particular characteristic of the product. The cost of destructive sampling often dictates small samples. For example, suppose a manufacturer of printers for personal computers wishes to estimate the mean number of characters printed before the printhead fails. Suppose the printer manufacturer tests n = 15 randomly selected printheads and records the number of characters printed until failure of each. These 15 measurements (in millions of characters) are listed in Table 5.1 below. Form a 99% confidence interval for the mean number of characters printed before the printhead fails.

Table 5.1

Number of Characters (In Millions)			
1.13	1.32	1.18	1.25
1.36	1.33	0.92	1.48
1.2	1.43	1.07	1.29
1.55	0.85	1.22	

Solution:

We solve Exercise 5.1 utilizing the **1 Vat t Interval** technique presented in DDXL. **Open** the Data File **PRINTHEAD** by following the directions found in the preface of this manual. If done correctly, the data should appear in a workbook similar to that shown below in Figure 5.3. Use the mouse to select the data shown in the workbook.

Figure 5.3

	A
1	Characters
2	1.13
3	1.55
4	1.43
5	0.92
6	1.25
7	1.36
8	1.32
9	0.85
10	1.07
11	1.48
12	1.2
13	1.33
14	1.18
15	1.22
16	1.29

Click on the **DDXL Add-In** menu. Click on the **Confidence Intervals** option to access the **Confidence Intervals** menu (see Figure 5.1 above). Click on the in the Function Type box to access the different confidence intervals available to select. Highlight the **1 Var t Interval** option to access the Confidence Interval Dialog menu shown in Figure 5.4. Highlight the **Characters** variable found in the **Names and Columns Box**. Click on the to the right of the **Quantitative Variable** box to create a confidence interval for the Characters variable. Click **OK**.

Figure 5.4

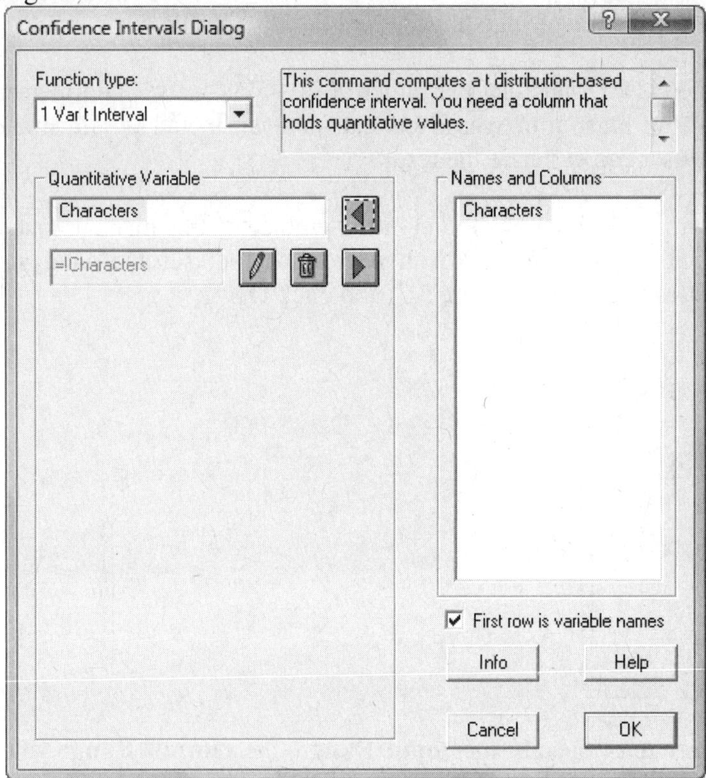

DDXL now asks you to select the level of reliability, or confidence, that you desire for the confidence interval. DDXL offers three levels of reliability to select – 90%, 95%, or 99%. We desire a 99% confidence interval in this problem, so we click on **99%** and then click on **Compute Interval**.

Figure 5.5

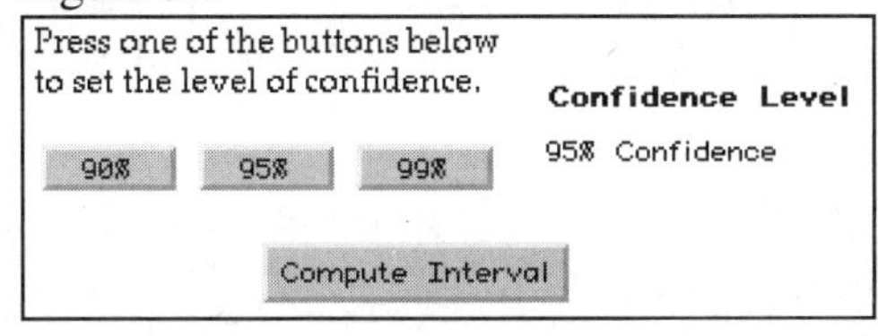

The resulting output is shown in Figure 5.6 below.

Figure 5.6

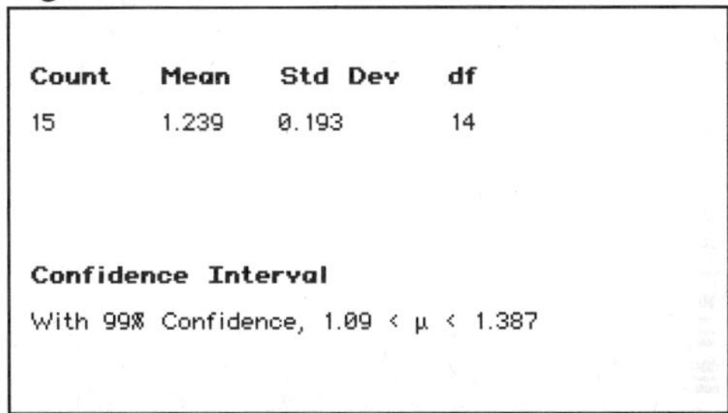

The DDXL output gives summary information and the computed confidence interval endpoints. In this example, the 99% confidence interval for the population mean is the interval (1.09, 1.387).

DDXL is very easy to use to generate confidence intervals that use either 90%, 95%, or 99% confidence levels. For other levels of reliability, however, we must utilize the Excel data analysis menu. We illustrate that technique using the same data so that the user can compare the results.

Open the Data File **PRINTHEAD** by following the directions found in the preface of this manual. Once the data is available, click on the **Data** tab and choose the **Data Analysis** option found in the **Analysis** group. Next, highlight the **Descriptive Statistics** option (see Figure 5.7) and click **OK**.

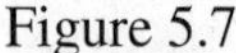

Figure 5.7

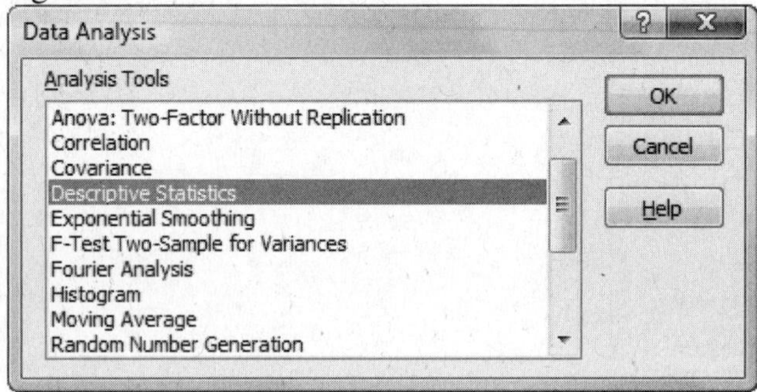

From the Descriptive Statistics menu, the user must specify the **Input Range**, the **Output Range**, and which statistics are desired. The Input Range indicates the worksheet location of the data to be analyzed. Either type or highlight with the mouse and enter the data set location for the Input Range (see Figure 5.8). The Output Range can either be a location within the current worksheet or a new Worksheet that you define. We opt to place the output in cell D1 of the current worksheet. As we saw in Chapter 2, the **Summary Statistics** box should be checked to provide the descriptive measures that were presented in Chapter 2. To generate a confidence interval for the data, check the **Confidence Level of Mean:** box and insert a level of reliability in the space next to the box. Click **OK**.

Figure 5.8

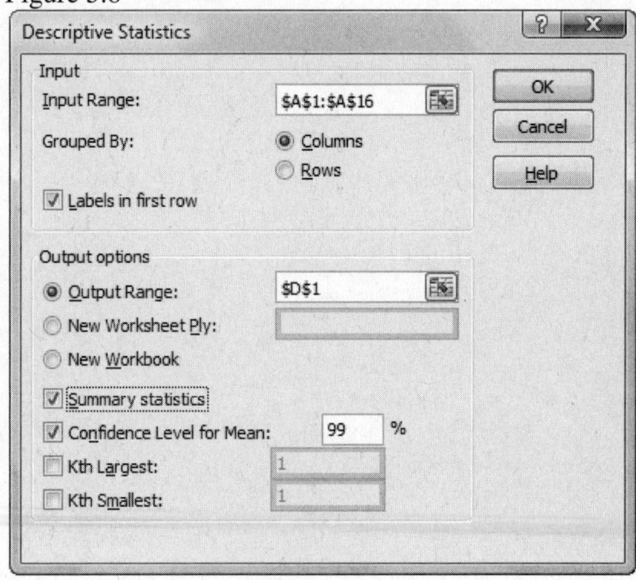

To attempt to match the results from the DDXL output, we use 99% as the confidence level and click on OK. The results are presented in Table 5.2 below:

Table 5.2

Characters	
Mean	1.238666667
Standard Error	0.049874764
Median	1.25
Mode	#N/A
Standard Deviation	0.19316413
Sample Variance	0.037312381
Kurtosis	0.063635947
Skewness	-0.491257803
Range	0.7
Minimum	0.85
Maximum	1.55
Sum	18.58
Count	15
Confidence Level(99.0%)	0.148469328

The confidence interval can be found by using $Mean \pm Confidence\ Level(99.0\%)$.

We find the resulting interval is (1.09, 1.387) which matches the interval we found using DDXL.
It is important to note that both the Excel Data Analysis Tool and the DDXL confidence interval provide identical results. While working with DDXL may be easier, the Excel Data Analysis Tool allows the user to specify any level of reliability that is desired. You are not restricted to the 90%, 95%, and 99% reliability levels.

5.3 Estimation of a Population Proportion

When estimating a population proportion, it is necessary for the user to identify one of the values as a success and the other value as a failure. DDXL asks the user to specify the success and then calculates a confidence interval for the proportion of successes. We illustrate using the next exercise.

Exercise 5.2: We use Example 5.7 from the *Statistics for Business and Economics* text.

Many public polling agencies conduct surveys to determine the current consumer sentiment concerning the state of the economy. For example, the Bureau of Economic and Business Research (BEBR) at the University of Florida conducts quarterly surveys to gauge consumer sentiment in the Sunshine State. Suppose that BEBR randomly samples 484 consumers and finds that 257 are optimistic about the state of the economy. Use a 90% confidence interval to estimate the proportion of all consumers in Florida who are optimistic about the state of the economy.

Solution:

Open the Data File **POLLING** by following the directions found in the preface of this manual. If done correctly, the data should appear in a workbook similar to that shown below in Figure 5.9. Use the mouse to select the data shown in the workbook.

Figure 5.9

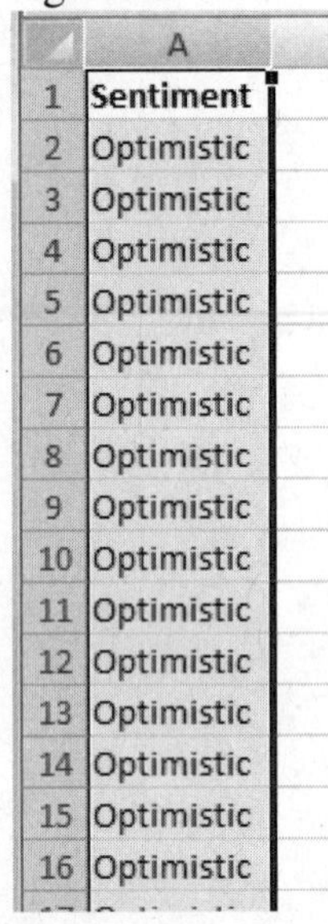

We should note that this data set was created to match the data given for this problem. We created the Sentiment variable and then entered 257 Optimistic responses and 227 Pessimistic responses. These 484 total responses were then saved in the POLLING data set. DDXL requires a data set to construct the confidence interval for a population proportion.

Click on the **DDXL Add-In** menu. Click on the **Confidence Intervals** option to access the **Confidence Intervals** menu (see Figure 5.1 above). Click on the ▾ in the Function Type box to access the different confidence intevals available to select. Highlight the **1 Var Prop Interval** option to access the Confidence Interval Dialog menu shown in Figure 5.4. Highlight the **Sentiment** variable found in the **Names and Columns Box**. Click on the ◀ to the right of the **Proportion Variable** box to create a confidence interval for the Characters variable. Click **OK**.

Figure 5.10

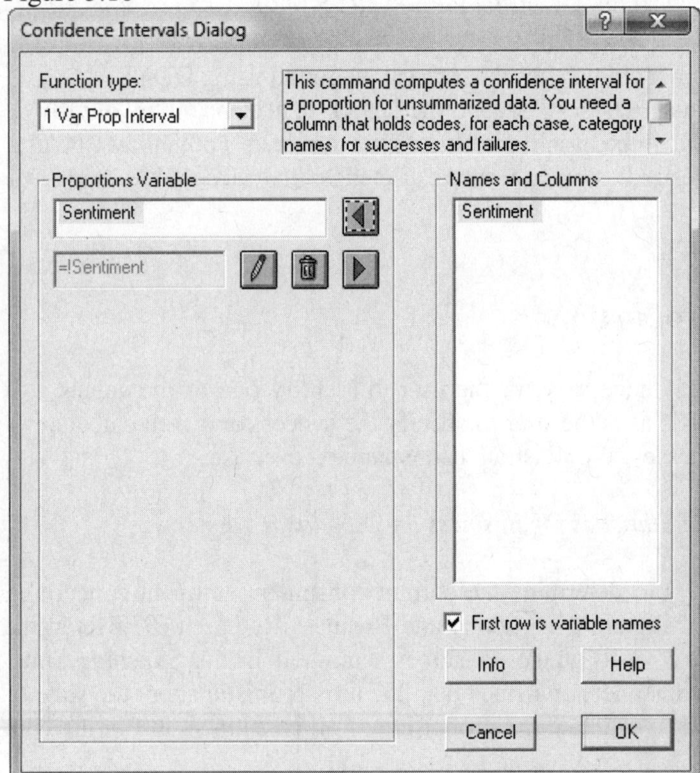

DDXL requires the user to make two choices prior to calculating the confidence interval desired. Step 1 requires the user to specify the condition that is being considered a success in the problem. By clicking on the **Set Success** button (shown in Figure 5.11),

Figure 5.11

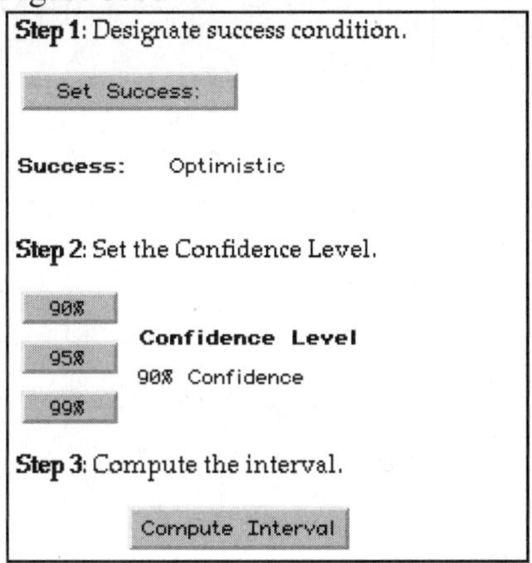

DDXL scrolls through all the responses that are given in the data set. The user should select the option that corresponds to the appropriate choice for a success. Because we are estimating the proportion of all Florida consumers who are optimistic about the economy, we should select **Optimist** as our choice of a **Success**.

In Step2, DDXL asks the user to select the level of reliability, or confidence, that you desire for the confidence interval. DDXL offers three levels of reliability to select – 90%, 95%, or 99%. We desire a 90% confidence interval in this problem, so we click on **90%** and then click on **Compute Interval**.

Figure 5.12

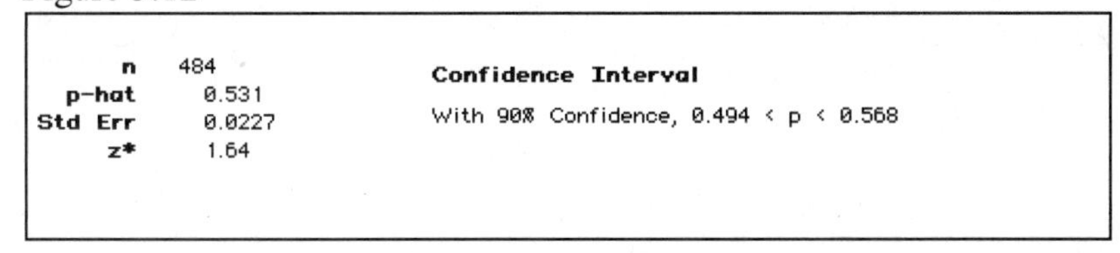

Figure 5.12 shows us the computed confidence interval. We see the proportion of all Florida consumers who are optimistic about the economy falls in the interval (.494, .568).

As we saw when creating a confidence interval for a population mean, DDXL is very easy to use to generate confidence intervals that use either 90%, 95%, or 99% confidence levels. For other levels of reliability, however, we must utilize the Excel data analysis menu. We illustrate that technique using the same data so that the user can compare the results.

Open the Data File **POLLING** by following the directions found in the preface of this manual. The Data Analysis option that we will be again utilizing within Excel does not allow the use of text variables. To work with this analysis tool, we must represent the text responses (Optimist and Pessimist) as either the value 0 or 1. We assign the value of 1 to the outcome we label as a success (in this case, Optimist) and the value of 0 to the outcome we label as a failure (in this case Pessimist). Once the data is available in 0 and 1 form, click on the **Data** tab and choose the **Data Analysis** option found in the **Analysis** group. Next, highlight the **Descriptive Statistics** option (see Figure 5.13) and click **OK**.

Figure 5.13

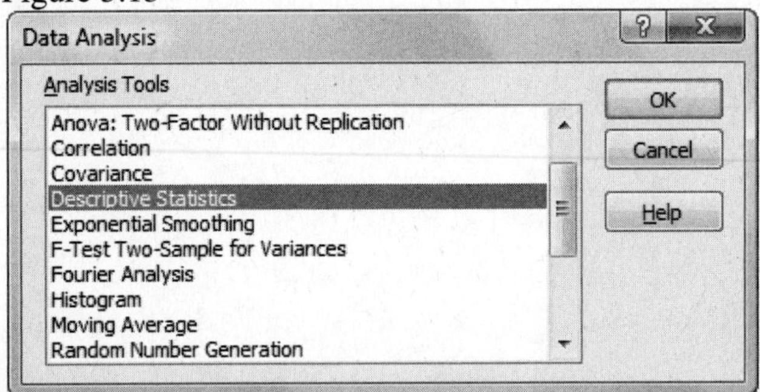

From the Descriptive Statistics menu, the user must specify the **Input Range**, the **Output Range**, and which statistics are desired. The Input Range indicates the worksheet location of the data to be analyzed. Either type or highlight with the mouse and enter the data set location for the Input Range (see Figure 5.14). The Output Range can either be a location within the current worksheet or a new Worksheet that you define. We opt to place the output in cell D1 of the current worksheet. As we saw in Chapter 2, the **Summary Statistics** box should be checked to provide the descriptive measures that were presented in Chapter 2. To generate a confidence interval for the data, check the **Confidence Level of Mean:** box and insert a level of reliability in the space next to the box. Click **OK**.

Figure 5.14

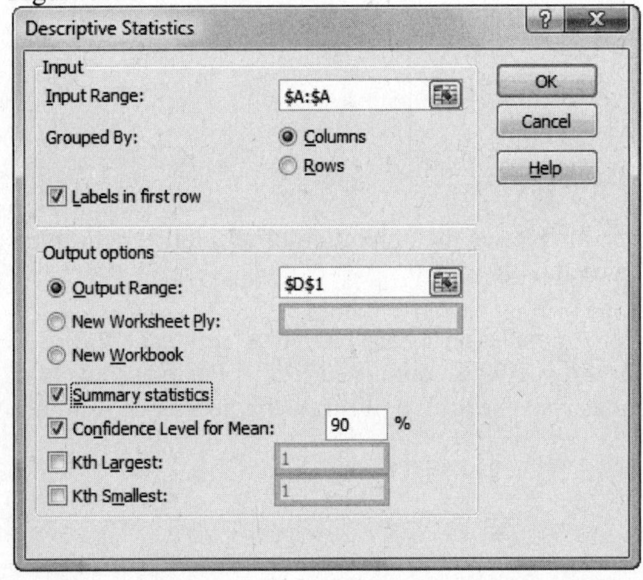

To attempt to match the results from the DDXL output, we use 90% as the confidence level and click on OK. The results are presented in Table 5.3 below:

Table 5.3

Sentiment	
Mean	0.530991736
Standard Error	0.022707042
Median	1
Mode	1
Standard Deviation	0.499554924
Sample Variance	0.249555122
Kurtosis	-1.992728371
Skewness	-0.124592234
Range	1
Minimum	0
Maximum	1
Sum	257
Count	484
Confidence Level(90.0%)	0.037421535

The confidence interval can be found by using $Mean \pm Confidence\ Level(90.0\%)$.

We find the resulting interval is (.494, .568), which matches the interval we found using DDXL.

It is important to note that both the Excel Data Analysis Tool and the DDXL confidence interval provide identical results. While working with DDXL may be easier, the Excel Data Analysis Tool allows the user to specify any level of reliability that is desired. You are not restricted to the 90%, 95%, and 99% reliability levels.

One final note about using Excel to work with proportions is needed. The Data Analysis tool within Excel always uses the *t*-distribution when constructing confidence intervals. When working with proportions, the *z*-distribution should be used in these calculations. For large sample sizes, the difference between the *z*- and the *t*-distributions is minimal. However, for smaller sample sizes, the difference between the two distributions can be substantial. For this reason, care should be taken when using the Data Analysis option to calculate confidence interval endpoints for estimating a population proportion.

Technology Lab

The following exercises from the *Statistics for Business and Economics* text are given for you to practice the confidence interval procedures that are available within DDXL. Included with the exercise is the DDXL output that was generated to solve the problem.

5.32 Methyl *t*-butyl ether (mtbe0 is an organic water contaminant that often results from gasoline spills. The level of MTBE (in parts per billion) was measured for a sample of 12 well sites located near a gasoline service station in New Jersey (*Environmental Science & Technology*, Jan. 2005). The data are listed in the accompanying table.

NJGAS

150	367	38	12	11	134
12	251	63	8	13	107

a. Give a point estimate for μ, the true mean MTBE level for all well sites located near the New Jersey gasoline service station.

b. Calculate and interpret a 99% confidence interval for μ.

DDXL/Excel Output

Count	Mean	Std Dev	df
12	97.167	113.76	11

Confidence Interval

With 99% Confidence, $-4.827 < \mu < 199.16$

MTBE	
Mean	97.16667
Standard Error	32.83956
Median	50.5
Mode	12
Standard Deviation	113.7596
Sample Variance	12941.24
Kurtosis	1.760952
Skewness	1.486316
Range	359
Minimum	8
Maximum	367
Sum	1166
Count	12
Confidence Level(99.0%)	101.9933

5.107 The primary determinant of the amount of vacation time U.S. employees receive is their length of service. According to data released by Hewitt Associates (*Management Review*, Nov. 1995), more than 8 of 10 employers provide 2 weeks of vacation after the first year. After 5 years, 75% of employers provide 3 weeks and after 15 years most provide 4-week vacations. To more accurately estimate p, the proportion of U.S. employers who provide only 2 weeks of vacation to new hires, a random sample of 24 major U.S. companies was contacted. The following vacation times were reported (in days).

VACTIMES

10	12	10	10	10	10
15	10	10	10	10	10
10	10	10	10	10	15
10	10	15	10	10	10

a. Construct a 95% confidence interval for *p*.

DDXL Output

n	24	**Confidence Interval**
p-hat	0.833	
Std Err	0.0761	With 95% Confidence, 0.684 < p < 0.982
z*	1.96	

Chapter 6
Inferences Based on a Single Sample: Tests of Hypothesis

6.1 Introduction

Chapter 6 introduces the reader to the concepts of hypothesis testing. The general theory and concepts of the test of hypothesis are then examined for inferences based on a single sample. Tests for a single population mean, a single population proportion, and a single population variance are all discussed in Chapter 6. In addition, the observed significance level of a test of hypothesis is explained and demonstrated in several examples.

DDXL provides procedures for all of the desired tests and computes p-values for all of these tests. DDXL provides two separate procedures for testing a population mean; when the population standard deviation is known (called the z-test in DDXL), and when the population standard deviation is unknown (called the t-test in DDXL). Since the population standard deviation is almost never known, we concentrate our work on the unknown standard deviation case. DDXL provides a test of a single population proportion that is calls the one proportion z-test. Finally, DDXL provides a chi-square test for a single population standard deviation that we will use to test a single population variance. We give examples of three of these procedures.

The following examples from *Statistics for Business and Economics* are solved using DDXL in this chapter:

Excel Companion		**Statistics for Business and Economics**	
Exercise	**Page**	**Example**	**Excel File Name**
6.1	100	Example 6.7	Emissions
6.2	104	Example 6.9	Defectives
6.3	107	Example 6.12	FillAmounts

6.2 Tests of Hypothesis of a Population Mean - Sigma Unknown

When testing a population mean, it is highly unlikely that the population standard deviation will be known. In such cases, it is necessary to estimate the value of the population standard deviation. We illustrate how DDXL can be used if such a test of hypothesis is desired.

To use the test of hypothesis tool within DDXL, **open** a new workbook and place the cursor in the upper left cell of the worksheet. Click on the **DDXL Add-In** menu. Click on the **Hypothesis Tests** option to access the **Hypothesis Tests** menu (see Figure 6.1 below). Click on the ▾ in the Function Type box to access the different tests available to select. Highlight the **1 Var t Test** option to access the Hypothesis Test Dialog menu shown in Figure 6.2.

Figure 6.1

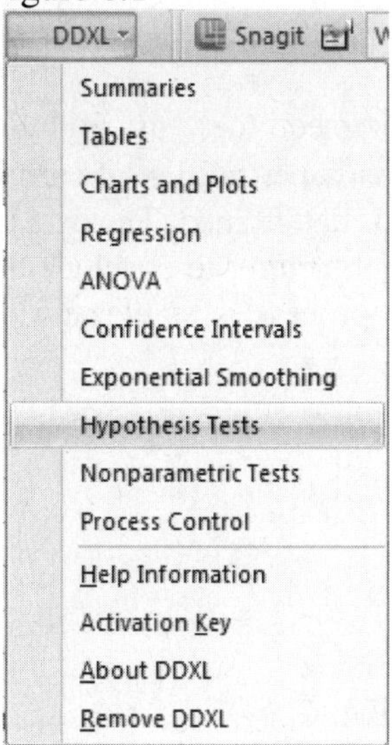

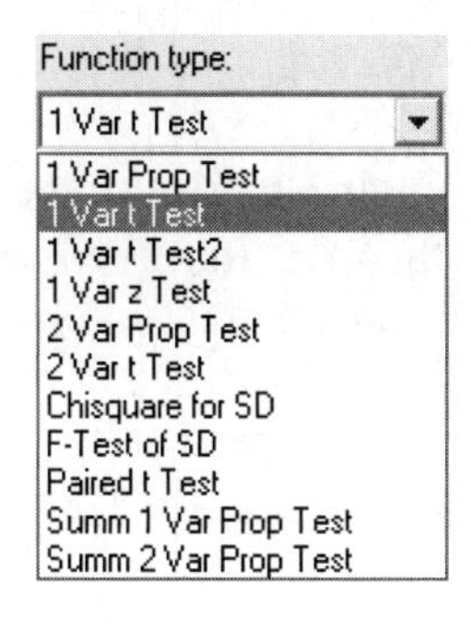

Figure 6.2

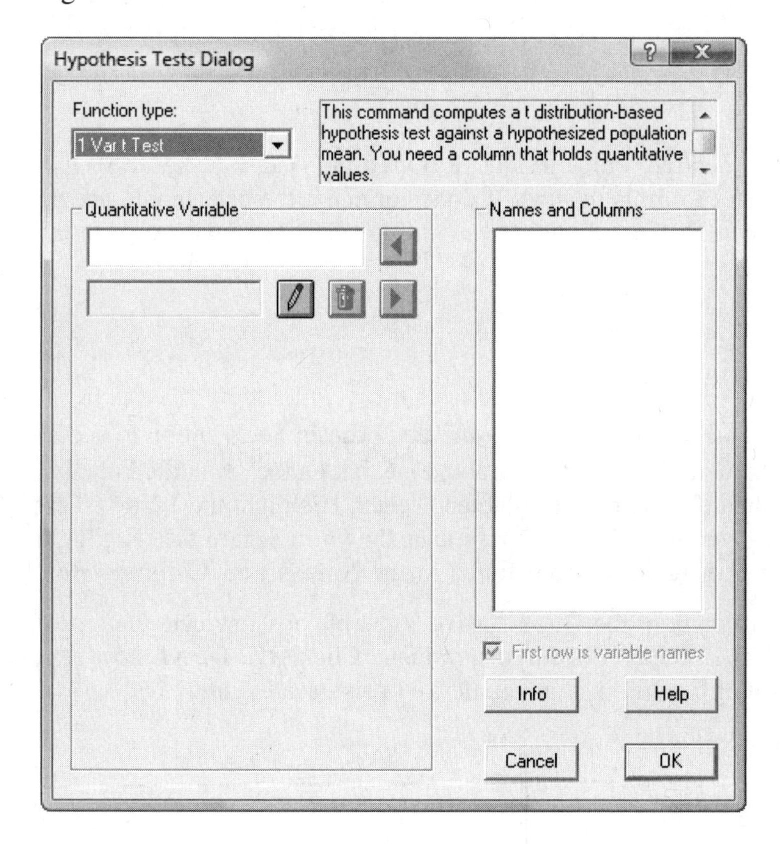

We illustrate how to use this technique with the following exercise:

Exercise 6.1: We use Example 6.7 found in the *Statistics for Business and Economics* text.

A major car manufacturer wants to test a new engine to determine whether it meets new air pollution standards. The mean emission μ of all engines of this type must be less than 20 parts per million of carbon. Ten engines are manufactured for testing purposes, and the emission level of each is determined. The data are shown in Table 6.1 below. Do the data supply sufficient evidence to allow the manufacturer to conclude that this type of engine meets the pollution standard? Assume that the production process is stable and the manufacturer is willing to risk a Type I error with probability $\alpha = .01$.

Table 6.1

Emission Level of Engine
15.6
16.2
22.5
20.5
16.4
19.4
19.6
17.9
12.7
14.9

Solution:

We solve Exercise 6.1 utilizing the **1 Vat t Test** presented in DDXL. **Open** the Data File **EMISSIONS** by following the directions found in the preface of this manual. If done correctly, the data should appear in a workbook similar to that shown below in Figure 6.3. Use the mouse to select the data shown in the workbook.

Figure 6.3

	A
1	Emission Level of Engine
2	15.6
3	16.2
4	22.5
5	20.5
6	16.4
7	19.4
8	19.6
9	17.9
10	12.7
11	14.9
12	

Click on the **DDXL Add-In** menu. Click on the **Hypothesis Tests** option to access the **Hypothesis Tests** menu (see Figure 6.1 above). Click on the ▾ in the Function Type box to access the different tests available to select. Highlight the **1 Var t Test** option to access the Hypothesis Test Dialog menu shown in Figure 6.4. Highlight the **Emission Level of Engine** variable found in the **Names and Columns Box**.

Click on the ◀ to the right of the **Quantitative Variable** box to create the test of hypothesis for the Emission Level of Engine variable. Click **OK**. DDXL now asks you to specify the test of hypothesis that you desire to test (see Figure 6.5).

Figure 6.4

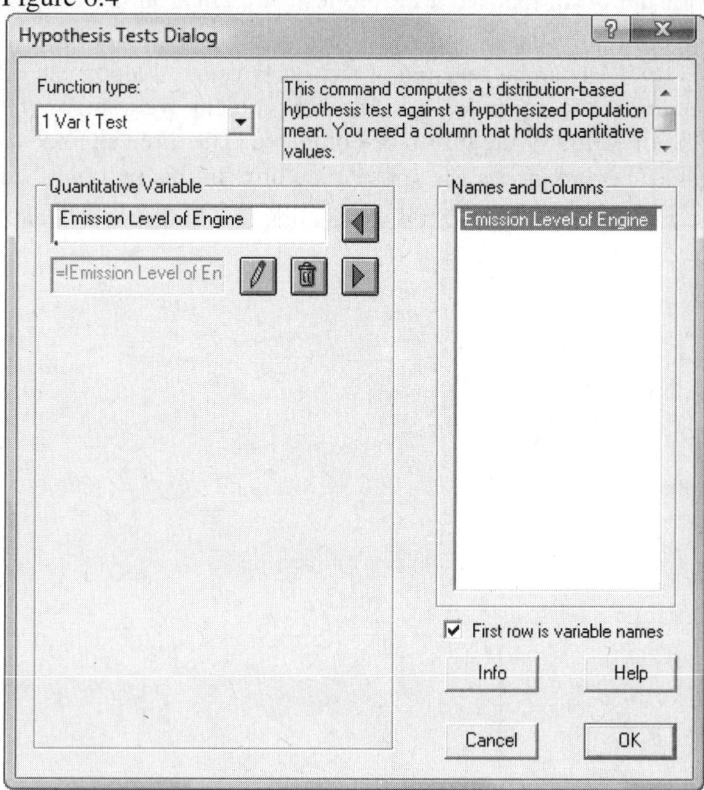

Figure 6.5

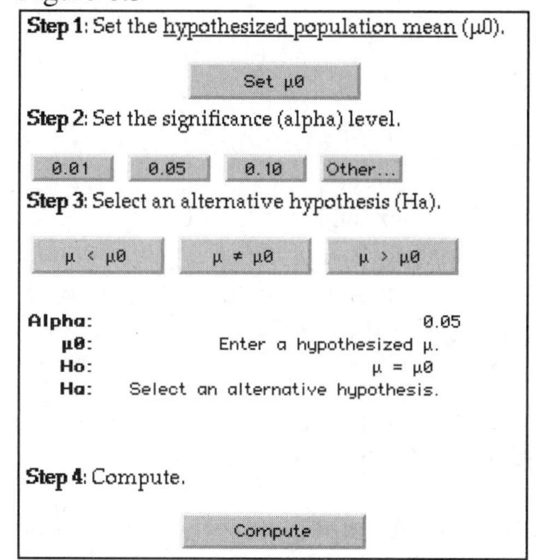

In **Step1**, we specify the hypothesized value of the mean that we desire to test. We click on the **Set** $\mu 0$ box to enter the value **20** for this problem (Figure 6.6). **Step 2** asks us to specify the alpha level that we desire to use for the test. Choices of **0.01**, **0.05**, or **0.10** can be selected or any other value of alpha can be entered by clicking on the **Other...** box and entering the desired value. For this problem, we click on the **0.01** box. **Step 3** requires the user to specify the direction of the test to be computed. The three alternative hypothesis choices $(<, >, and \neq)$ can be selected by clicking on the appropriate box. In this problem, we select a lower tail test by clicking on the $\mu < \mu 0$ box. Finally, in **Step 4**, we click on **Compute** to create the output shown in Figure 6.7.

Figure 6.6

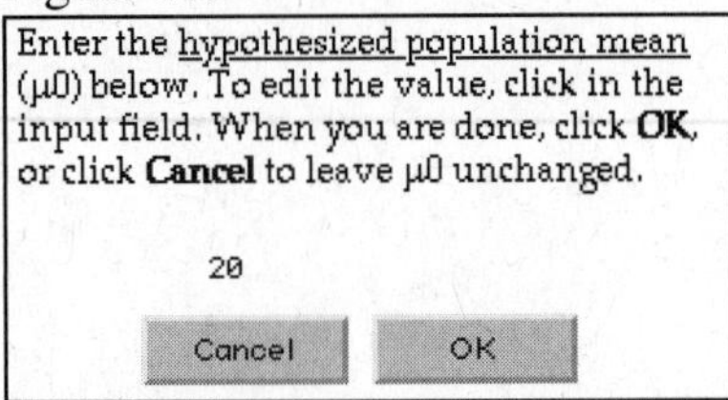

Figure 6.7

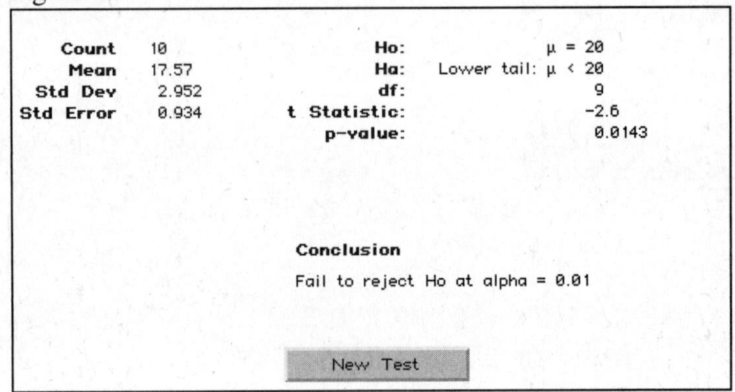

We see from the printout that the computed test statistic is t = -2.6, the p-value for the test is p = .0143, and the appropriate conclusion for the test is fail to reject Ho when testing at alpha = .01. We compare these results with the ones given in the text to verify that we are conducting the test of hypothesis correctly.

6.3 Tests of Hypothesis of a Population Proportion

To use the test of hypothesis tool for population proportions within DDXL, **open** a new workbook and place the cursor in the upper left cell of the worksheet. Click on the **DDXL Add-In** menu. Click on the **Hypothesis Tests** option to access the **Hypothesis Tests** menu (see Figure 6.8). Click on the ▾ in the Function Type box to access the different tests available to select. Highlight the **1 Var Prop Test** option to access the Hypothesis Test Dialog menu shown in Figure 6.9.

Figure 6.8

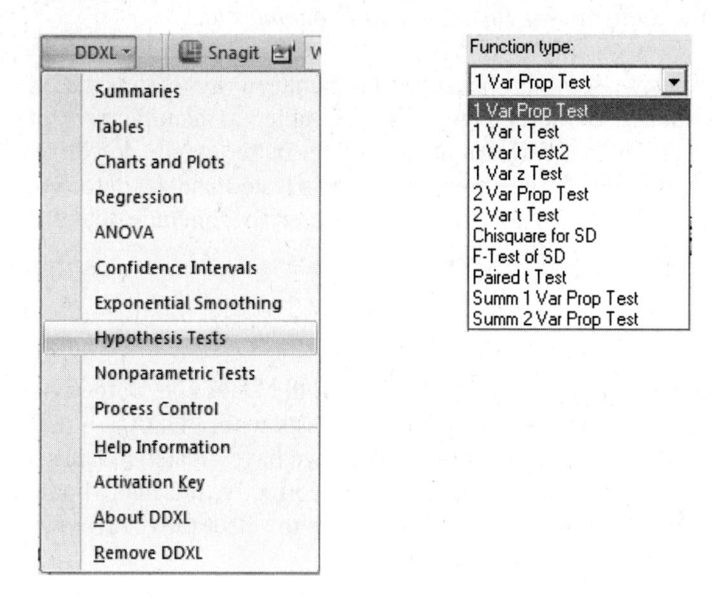

Figure 6.9

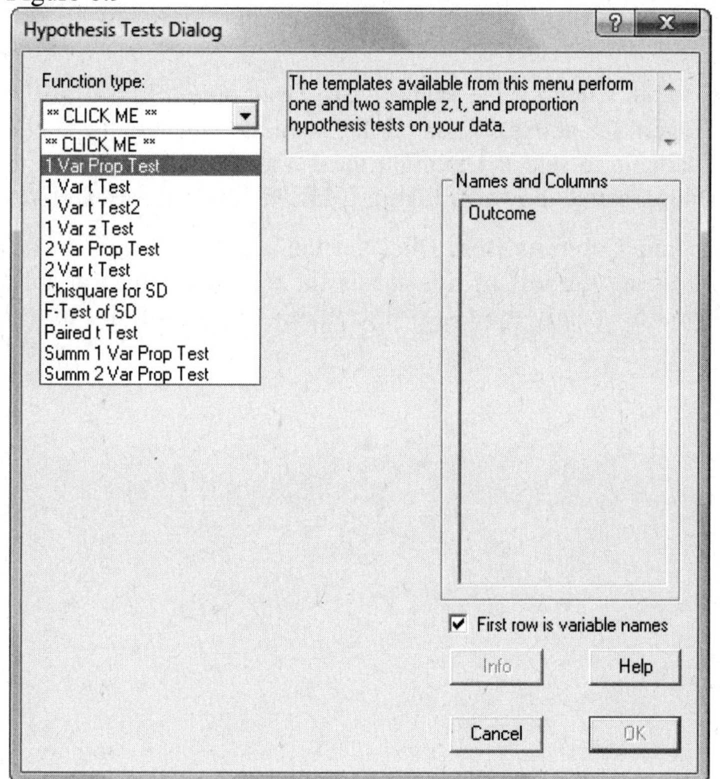

We illustrate how to use this technique with the following exercise:

Exercise 6.2: We use Example 6.9 found in the *Statistics for Business and Economics* text.

The reputations (and hence sales) of many businesses can be severely damaged by shipments of manufactured items that contain a large percentage of defectives. For example, a manufacturer of alkaline batteries may want to be reasonably certain that fewer than 5% of its batteries are defective. Suppose 300 batteries are randomly selected from a very large shipment; each is tested and 10 defective batteries are found. Does this provide sufficient evidence for the manufacturer to conclude that the fraction defective in the entire shipment is less than .05? Use $\alpha = .01$.

Solution:

We solve Exercise 6.2 utilizing the **1 Vat Prop Test** presented in DDXL. **Open** the Data File **Defectives** by following the directions found in the preface of this manual. For proportion tests within DDXL, a data set must be used that looks at the outcomes of an experiment. For this problem, we have created a data set that contains 10 defective batteries and 290 batteries that tested fine. If done correctly, the data should appear in a workbook similar to that shown below in Figure 6.10. Use the mouse to select the data shown in the workbook.

Figure 6.10

	A
1	Defective
2	Yes
3	Yes
4	Yes
5	Yes
6	Yes
7	Yes
8	Yes
9	Yes
10	Yes
11	Yes
12	No
13	No
14	No

Click on the **DDXL Add-In** menu. Click on the **Hypothesis Tests** option to access the **Hypothesis Tests** menu (see Figure 6.8 above). Click on the ▾ in the Function Type box to access the different tests available to select. Highlight the **1 Var Prop Test** option to access the Hypothesis Test Dialog menu shown in Figure 6.11. Highlight the **Defectives** variable found in the **Names and Columns Box**. Click on the ◀ to the right of the **Proportions Variable** box to create the test of hypothesis for the Defectives variable. Click **OK**. DDXL now asks you to specify the test of hypothesis that you desire to test (see Figure 6.12).

Figure 6.11

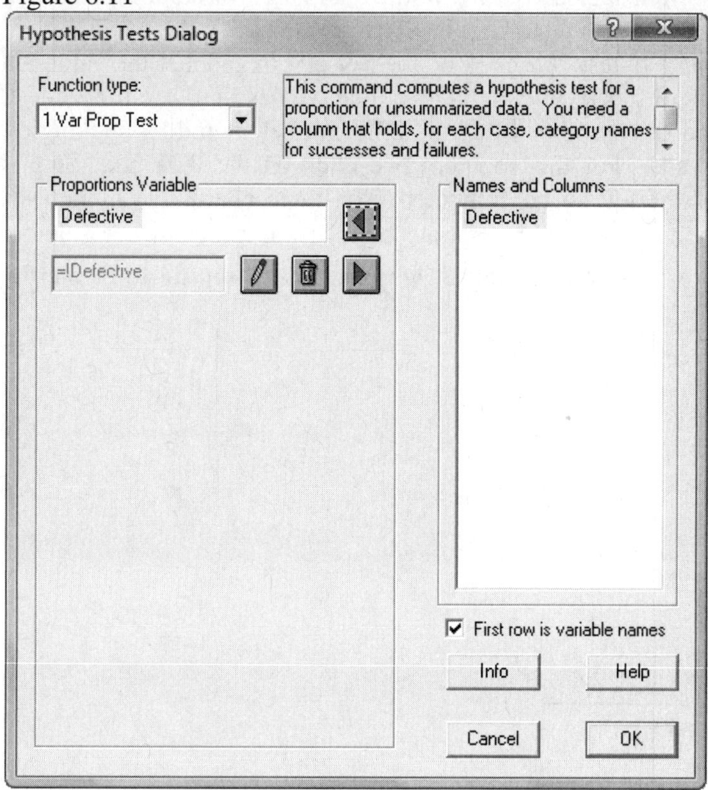

Figure 6.12

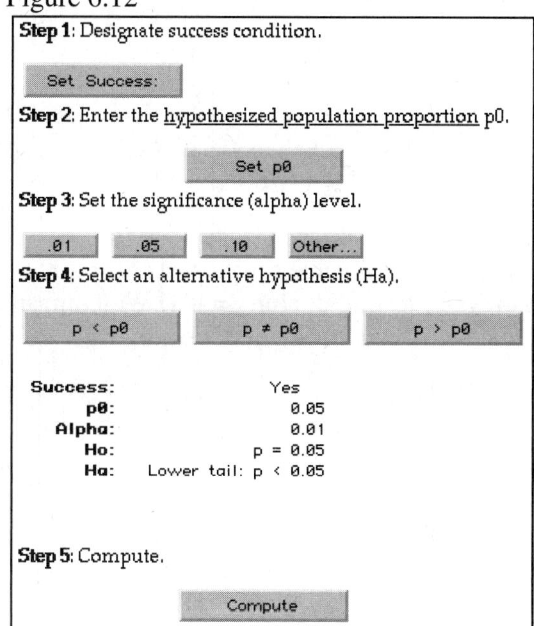

In **Step1**, we click on the **Set Success:** box until the correct setting of the qualitative variable is listed as a success. For this problem, we specify the value **Yes** as a success In **Step 2**, we specify the hypothesized value of the population proportion that we desire to test. We click on the **Set p0** box to enter the value **.05** for this problem (Figure 6.13). **Step 3** asks us to specify the alpha level that we desire to use for the test. Choices of **0.01**, **0.05**, or **0.10** can be selected or any other value of alpha can be entered by clicking on the **Other...** box and entering the desired value. For this problem, we click on the **0.01** box. **Step 4** requires the user to specify the direction of the test to be computed. The three alternative hypothesis choices $(<, >, and \neq)$ can be selected by clicking on the appropriate box. In this problem, we select a lower tail test by clicking on the $p < p0$ box. Finally, in **Step 5**, we click on **Compute** to create the output shown in Figure 6.14.

Figure 6.13

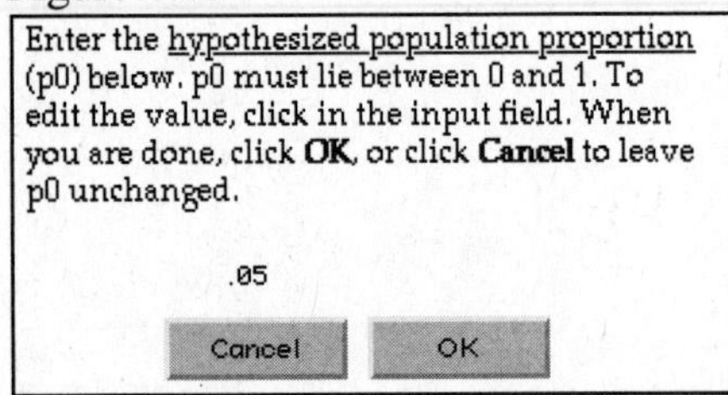

Figure 6.14

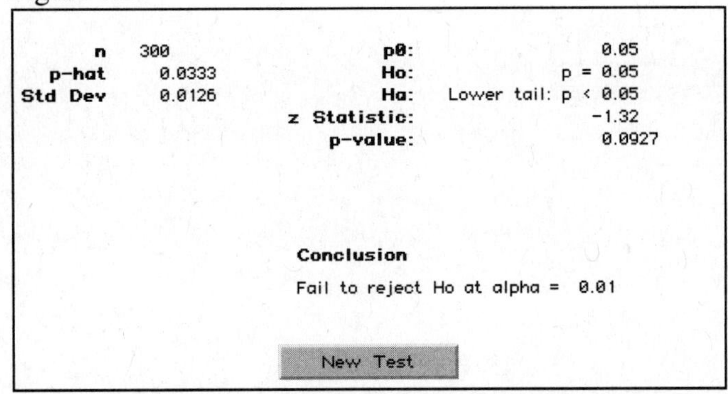

We see from the printout that the computed test statistic is t = -1.32, the p-value for the test is p = .0927, and the appropriate conclusion for the test is fail to reject Ho when testing at alpha = .01. We compare these results with the ones given in the text to verify that we are conducting the test of hypothesis correctly.

6.4 Tests of Hypothesis of a Population Standard Deviation

To use the test of hypothesis tool for population standard deviations within DDXL, **open** a new workbook and place the cursor in the upper left cell of the worksheet. Click on the **DDXL Add-In** menu. Click on the **Hypothesis Tests** option to access the **Hypothesis Tests** menu (see Figure 6.15). Click on the ▾ in the Function Type box to access the different tests available to select. Highlight the **1 Var Prop Test** option to access the Hypothesis Test Dialog menu shown in Figure 6.16 on the next page.

Figure 6.15

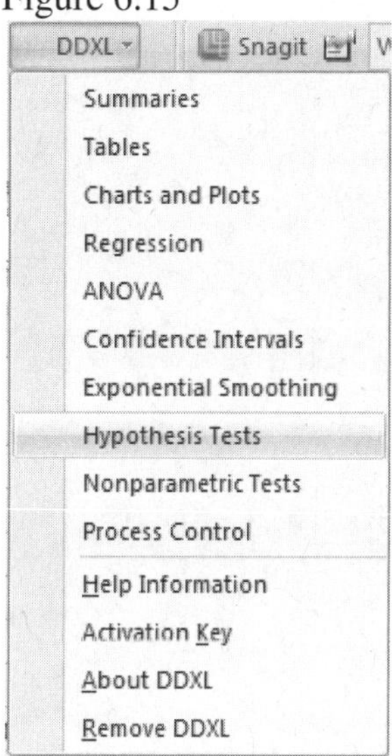

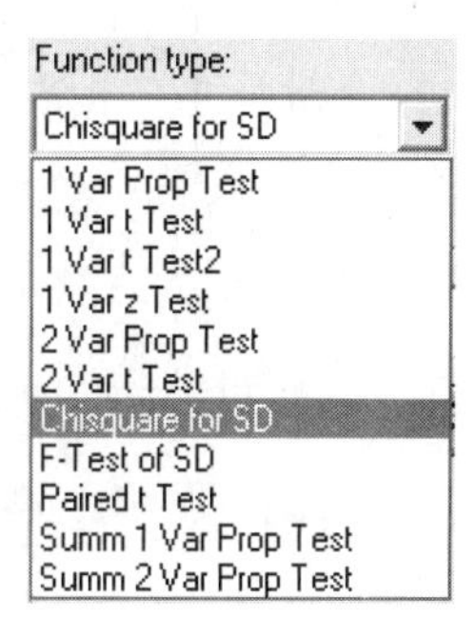

We illustrate how to use this technique with the following exercise:

Exercise 6.3: We use Example 6.12 found in the *Statistics for Business and Economics* text.

Refer to the fill weights for the sample of ten 16-ounce cans in Table 6.2. Do the data provide sufficient evidence to indicate that the true standard deviation σ of the fill measurements of all 16-ounce cans is less than .1 ounce?

Table 6.2

16.00	16.06	15.95	16.04	16.10	16.05	16.02	16.03	15.99	16.02

Solution:

We solve Exercise 6.3 utilizing the **Chisquare for Sd** test presented in DDXL. **Open** the Data File **FILLAMOUNTS** by following the directions found in the preface of this manual. If done correctly, the data should appear in a workbook similar to that shown below in Figure 6.17. Use the mouse to select the data shown in the workbook.

Figure 6.16

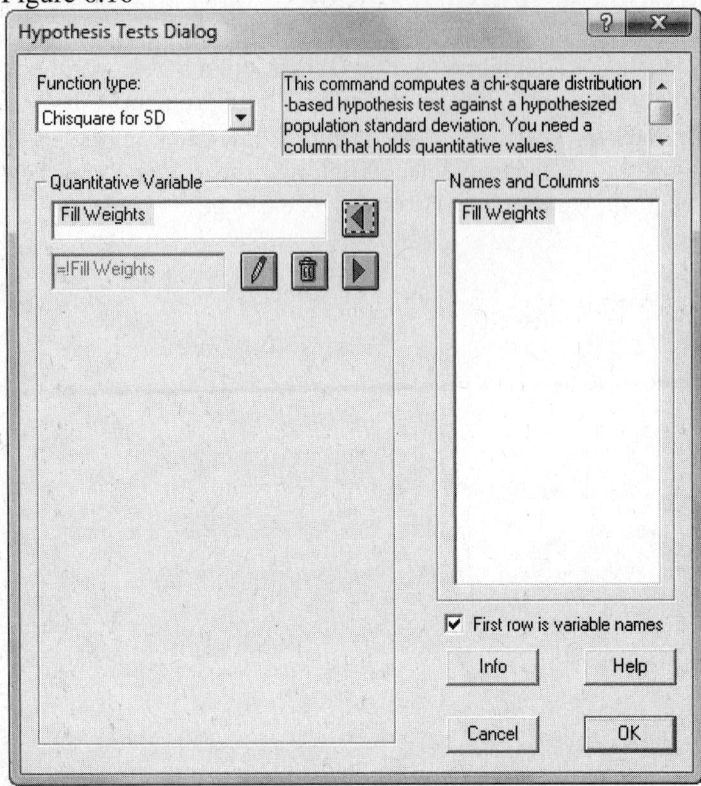

Figure 6.17

	A
1	Fill Weights
2	16
3	16.06
4	15.95
5	16.04
6	16.1
7	16.05
8	16.02
9	16.03
10	15.99
11	16.02

Click on the **DDXL Add-In** menu. Click on the **Hypothesis Tests** option to access the **Hypothesis Tests** menu (see Figure 6.15 above). Click on the ▼ in the Function Type box to access the different tests available to select. Highlight the **Chisquare for SD** test option to access the Hypothesis Test Dialog menu shown in Figure 6.16. Highlight the **Fill Weights** variable found in the **Names and Columns Box**. Click on the ◀ to the right of the **Quantitative Variable** box to create the test of hypothesis for the Fill Weights variable. Click **OK**. DDXL now asks you to specify the test of hypothesis that you desire to test (see Figure 6.18).

Figure 6.18

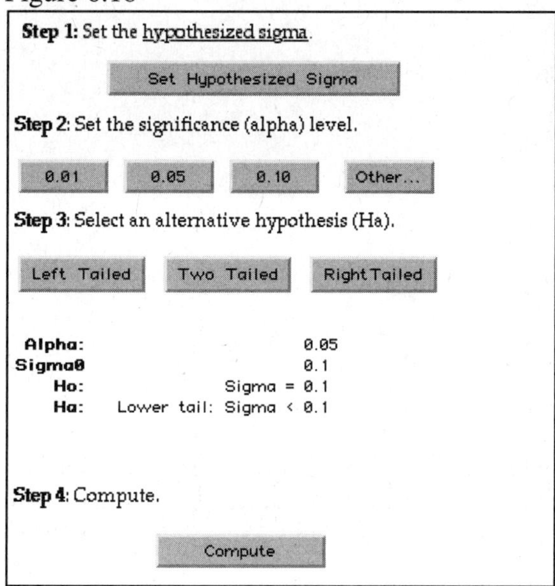

In **Step1**, we specify the hypothesized value of the standard deviation that we desire to test. We click on the **Set Hypothesized Sigma** box to enter the value **.10** for this problem (Figure 6.19). **Step 2** asks us to specify the alpha level that we desire to use for the test. Choices of **0.01**, **0.05**, or **0.10** can be selected or any other value of alpha can be entered by clicking on the **Other...** box and entering the desired value. For this problem, we click on the **0.05** box since that is the value of alpha that is used in the solution to the problem in the text. **Step 3** requires the user to specify the direction of the test to be computed. The three alternative hypothesis choices $(<, >, and \neq)$ can be selected by clicking on the appropriate box. In this problem, we select a lower tail test by clicking on the **Left Tailed** box. Finally, in **Step 4**, we click on **Compute** to create the output shown in Figure 6.20.

Figure 6.19

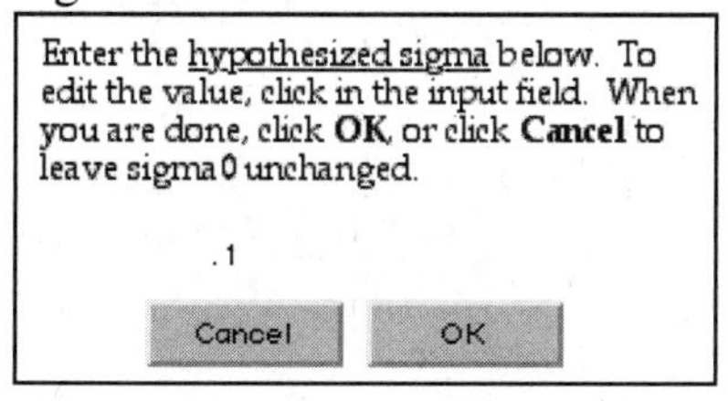

Figure 6.20

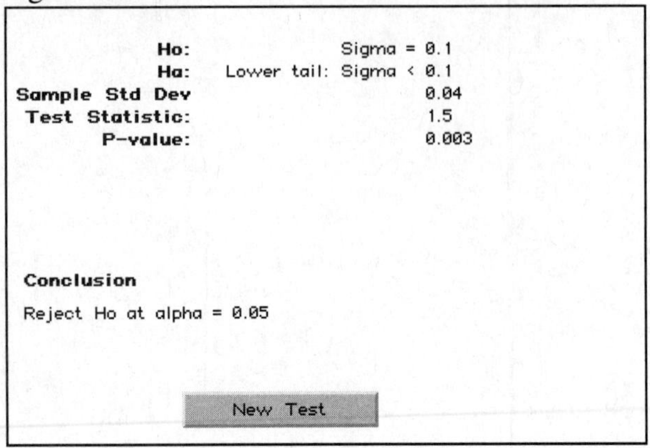
Ho: Sigma = 0.1
Ha: Lower tail: Sigma < 0.1
Sample Std Dev 0.04
Test Statistic: 1.5
P-value: 0.003

Conclusion

Reject Ho at alpha = 0.05

New Test

We see from the printout that the computed test statistic is Chi-Square = 1.5, the p-value for the test is p = .0003, and the appropriate conclusion for the test is to reject Ho when testing at alpha = .05. We compare these results with the ones given in the text to verify that we are conducting the test of hypothesis correctly.

Technology Lab

The following exercises from the *Statistics for Business and Economics* text are given for you to practice the procedures covered in the text that are available within DDXL. Included with the exercises are the DDXL outputs that were generated to solve the problems.

6.130 One way of evaluating a measuring instrument is to repeatedly measure the same item and compare the average of these measurements to the item's known measured value. The difference is used to assess the instrument's accuracy (*American Society for Quality*). To evaluate a particular Metlar scale, an item whose weight is known to be 16.01 ounces is weighed five times by the same operator. The measurements, in ounces, are as follows:

15.99	16.00	15.97	16.01	15.96

a. In a statistical sense, does the average measurement differ from 16.01? Conduct the appropriate hypothesis test. What does your analysis suggest about the accuracy of the instrument?

b. Evaluate the instrument's precision by testing whether the standard deviation of the weight measurements is greater than .01. Use α=.05.

DDXL Output 6.130 a

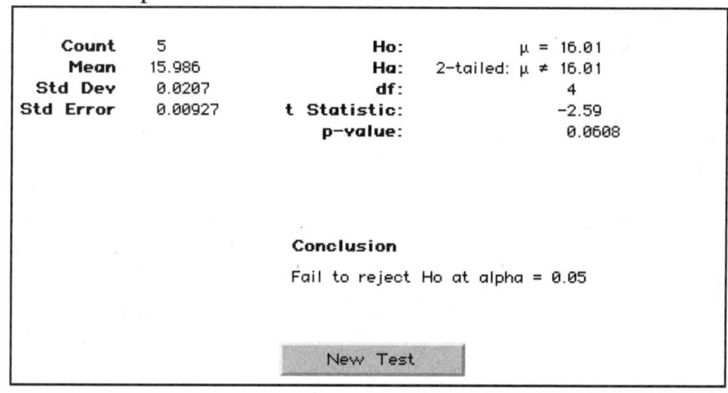

DDXL Output 6.130 c

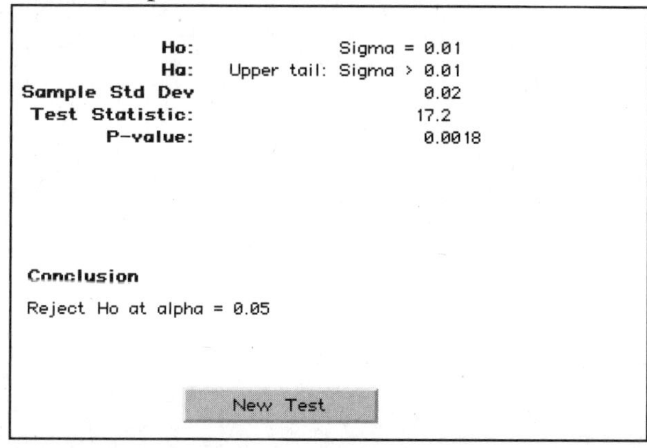

6.124 Shoplifting in the United States costs retailers about \$35 million a day. Despite the seriousness of the problem, the National Association of Shoplifting Prevention (NASP) claims that only 50% of all shoplifters are turned over to police (*www.shopliftingprevention.org*, 2009). A random sample of 40 U.S. retailers were questioned concerning the disposition of the most recent shoplifter they apprehended. A total of 24 were turned over to police. Do these data provide sufficient evidence to contradict the NASP?

DDXL Output

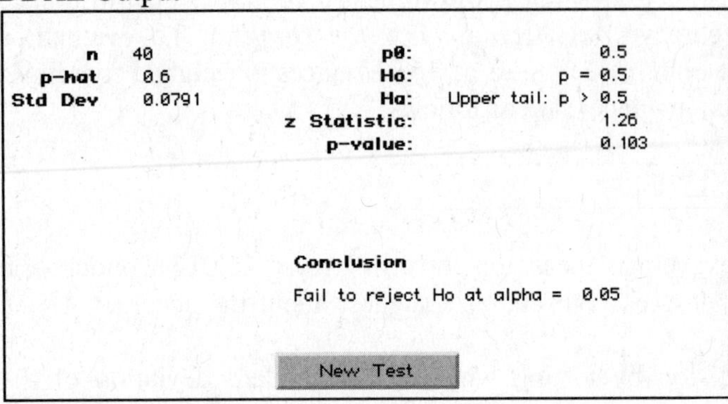

Chapter 7
Inferences Based on Two Samples: Confidence Intervals and Tests of Hypothesis

7.1 Introduction

Chapter 7 introduces the reader to two sample problems using both the estimation and test of hypothesis techniques discussed in Chapters 5 and 6. Three types of parameters, population means, population proportions, and population variances, are studied in the chapter. The reader is also introduced to the topic of sample size determination, as it follows very nicely from the estimation material presented.

DDXL provides calculation of both confidence intervals and tests of hypotheses for comparing population means, population proportions, and population standard deviations. DDXL provides both an independent comparison of population means and also a paired difference comparison of population means. DDXL does not provide any sample size determination techniques, but the formulas presented in the text are fairly easy to use.

DDXL requires the user to have files containing the data that is desired to be analyzed. When working with proportions, DDXL allows the user to enter summary information into the menus to compute the desired confidence interval or test of hypothesis. The following examples from *Statistics for Business and Economics* are solved using DDXL in this chapter:

Excel Companion Exercise	Page	Statistics for Business and Economics Example	Excel File Name
7.1	114	Example 7.1	AUTOSTUDY
7.2	117	Example 7.2	AUTOSTUDY
7.3	120	Example 7.5	Gradpairs
7.4	122	Example 7.6	Proportion
7.5	125	Example 7.10	PAPERMILLS

7.2 Comparing Two Means – Independent Sampling

The *Statistics for Business and Economics* text offers two techniques for comparing two population means, the independent sampling and matched pairs sampling techniques. In this section, we will look at how DDXL analyzes data collecting using two random, independent samples. To use these tools within DDXL, **open** a new workbook and highlight the data that is to be compared. Click on the **DDXL Add-In** menu. Click on **Confidence Intervals** or **Hypothesis Tests** option to access the desired menu. Click on the ▾ in the Function Type box to access the different procedures that are available for use. The option that would be selected for the independent comparison of means would be the **2 Var t Test** option shown in Figure 7.1.

Figure 7.1

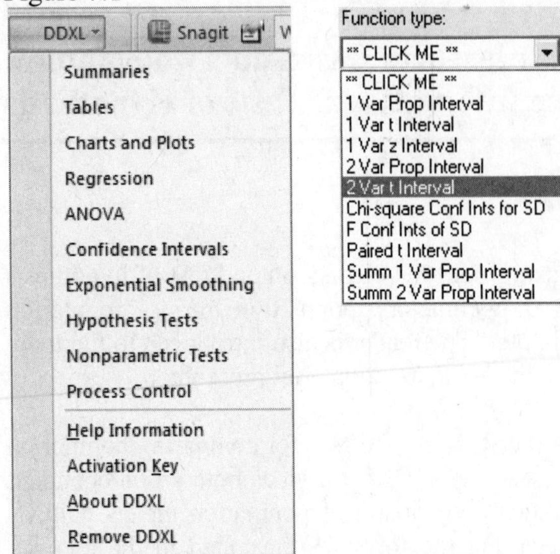

We illustrate how to use these procedures in the following two examples.

Exercise 7.1: We use Example 7.1 found in the *Statistics for Business and Economics* text.

In recent years, the United States and Japan have engaged in intense negotiations regarding restrictions on trade between the two countries. One of the claims made repeatedly by U.S. officials is that many Japanese manufacturers price their goods higher in Japan than in the United States, in effect subsidizing low prices in the United States by extremely high prices in Japan. According to the U.S. argument, Japan accomplishes this by keeping competitive U.S. goods from reaching the Japanese marketplace.

An economist decided to test the hypothesis that higher retail prices are being charged for Japanese automobiles in Japan than in the United States. She obtained random samples of 50 retail sales in the United States and 30 retail sales in Japan over the same time period and for the same model of automobile, converted the Japanese sales prices from yen to dollars using current conversion rates. The data, saved in the **AUTOSTUDY** file, are listed in Table 7.1. Form a 95% confidence interval for the difference between the population mean retail prices of this automobile model for the two countries. Interpret the results.

Table 7.1

USA	18200	16200	17200	18700	18400	16600	14900	16800	12100	10800
	18500	15500	16200	16300	18200	19500	13200	16800	12900	17200
	18200	16300	16800	16400	18600	15600	17100	18100	18900	19000
	17300	18800	14900	16700	20300	17100	14600	17200	13000	18400
	16900	13300	16300	15900	16600	17600	16000	17100	14600	18000
JAPAN	18500	14000	18200	21100	13900	18700	14900	16400	16300	18000
	16800	19800	17300	16600	14900	16300	16500	15400	17600	20100
	16400	18000	17500	18400	19800	14800	18200	16700	20200	16200
	20400	17900	15500	15400	17700	17100	17900	17400	18200	16200
	18500	16900	17600	14400	21600	18600	16200	14300	12500	20000

Solution:

We solve Exercise 7.1 utilizing the **2 Var t Test** presented in the DDXL **Confidence Intervals** menu. **Open** the Data File **AUTOSTUDY** by following the directions found in the preface of this manual. If done correctly, the data should appear in a workbook similar to that shown below in Figure 7.2. Use the mouse to select the data shown in the workbook.

Figure 7.2

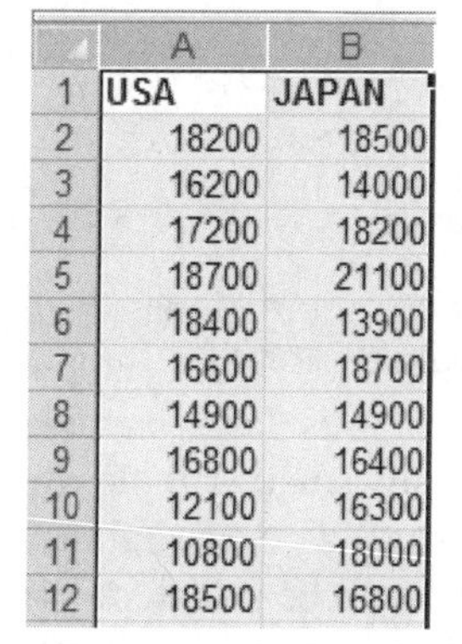

	A	B
1	USA	JAPAN
2	18200	18500
3	16200	14000
4	17200	18200
5	18700	21100
6	18400	13900
7	16600	18700
8	14900	14900
9	16800	16400
10	12100	16300
11	10800	18000
12	18500	16800

Click on the **DDXL Add-In** menu. Click on the **Confidence Intervals** option to access the **Confidence Intervals** menu. Click on the ▾ in the Function Type box to access the different procedures available to select. Highlight the **2 Var t Test** option to access the Confidence Interval Dialog menu shown in Figure 7.3. Highlight the **USA** variable found in the **Names and Columns Box** and click on the ◀ to the right of the **1st Quantitative Variable** box. Highlight the **Japan** variable found in the **Names and Columns Box** and click on the ◀ to the right of the **2nd Quantitative Variable** box. Click **OK**. DDXL now asks you to specify the confidence interval that you wish to create (see Figure 7.4).

Figure 7.3

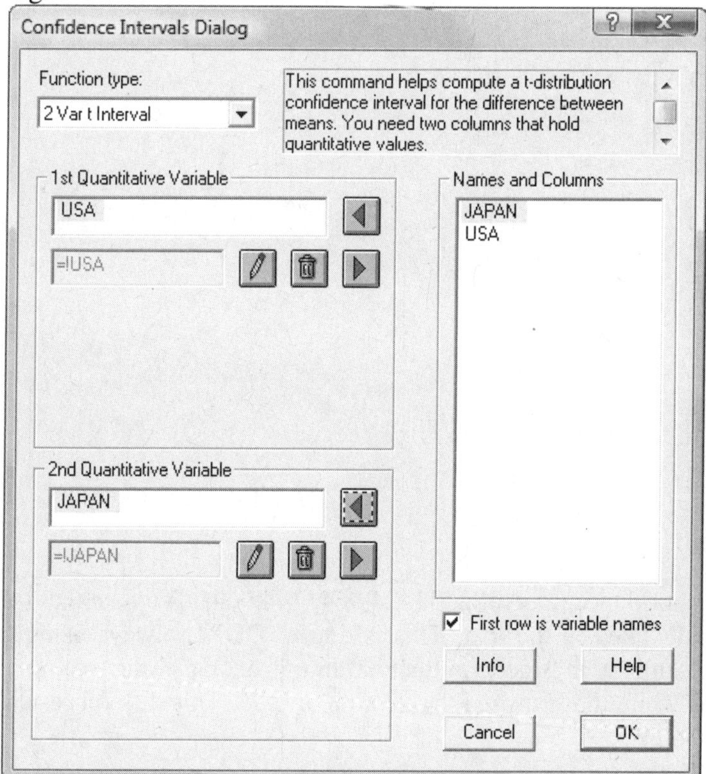

Figure 7.4

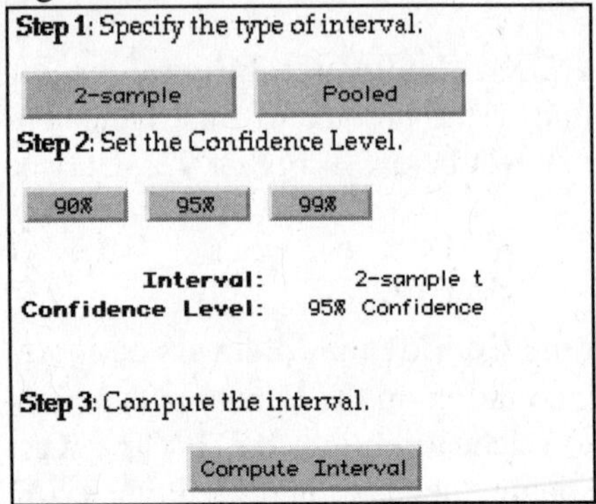

In **Step 1**, we specify how we want DDXL to estimate the variances of the two populations. Two techniques, discussed in the text, can be used. We click on the **2-sample** option as the individual sample variances should be used to estimate their corresponding population variances in this large sample comparison of means. The **Pooled** option would be appropriate when working with small samples in which the assumption of equal population variances is appropriate. **Step 2** asks us to specify the reliability level that we desire to use for the confidence interval. Choices of **90%, 95%,** or **99%** can be selected. For this problem, we click on the **95%** box. Finally, in **Step 3,** we click on **Compute** to create the output shown in Figure 7.5.

Figure 7.5

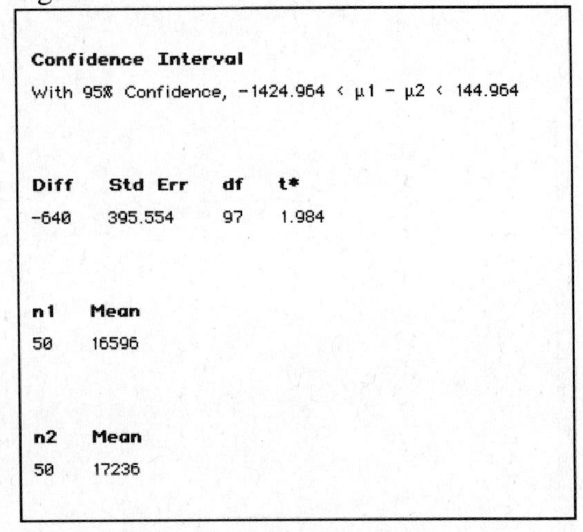

We see from the printout that the computed confidence interval is (-1424.964, 144.964). We note that the numbers are slightly different from those calculated in the text. This is due to DDXL always using t-values instead of using the z-values for large samples as is done in the text. In this example, the t-value of 1.984 was used by DDXL in the calculations while the text used the z-value of 1.96. This difference will cause slight discrepancies in the interval endpoints.

The next example illustrates how to use DDXL to conduct a test of hypothesis using two random and independent samples.

Exercise 7.2: We use Example 7.2 found in the *Statistics for Business and Economics* text.

Refer to the study of retail prices of an automobile sold in the United States and Japan, Example 7.2. Another way to compare the mean retail prices for the two countries is to conduct a test of hypothesis. Use DDXL to conduct the test. Use $\alpha = .05$.

Solution:

We solve Exercise 7.2 utilizing the **2 Var t Test** presented in the DDXL **Hypothesis Tests** menu. **Open** the Data File **AUTOSTUDY** by following the directions found in the preface of this manual. If done correctly, the data should appear in a workbook similar to that shown below in Figure 7.6. Use the mouse to select the data shown in the workbook.

Figure 7.6

	A	B
1	USA	JAPAN
2	18200	18500
3	16200	14000
4	17200	18200
5	18700	21100
6	18400	13900
7	16600	18700
8	14900	14900
9	16800	16400
10	12100	16300
11	10800	18000
12	18500	16800

Click on the **DDXL Add-In** menu. Click on the **Hypothesis Tests** option to access the **Hypothesis Tests** menu. Click on the ▾ in the Function Type box to access the different procedures available to select. Highlight the **2 Var t Test** option to access the Hypothesis Tests Dialog menu shown in Figure 7.7. Highlight the **USA** variable found in the **Names and Columns Box** and click on the ◀ to the right of the 1st **Quantitative Variable** box. Highlight the **Japan** variable found in the **Names and Columns Box** and click on the ◀ to the right of the 2nd **Quantitative Variable** box. Click **OK**. DDXL now asks you to specify the confidence interval that you wish to create (see Figure 7.8).

Figure 7.7

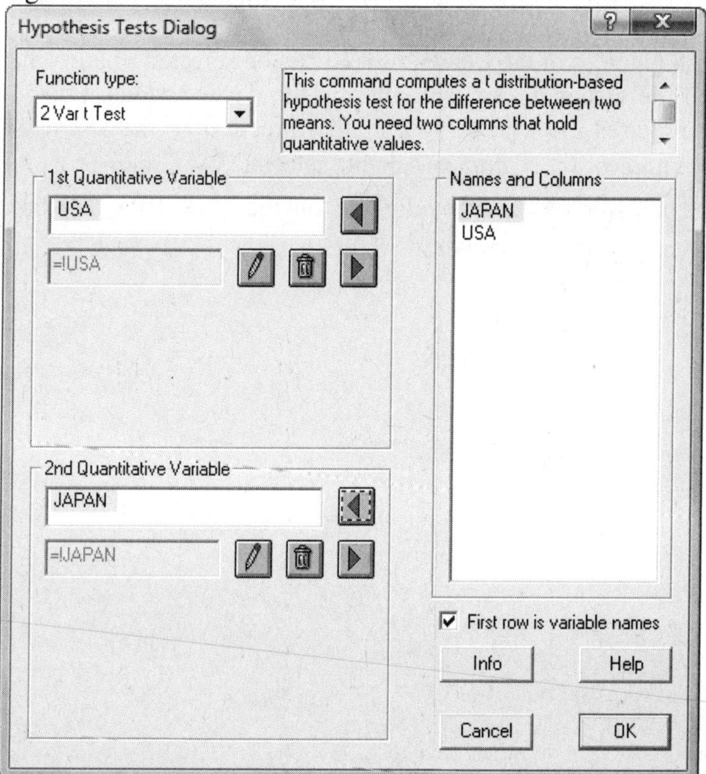

Figure 7.8

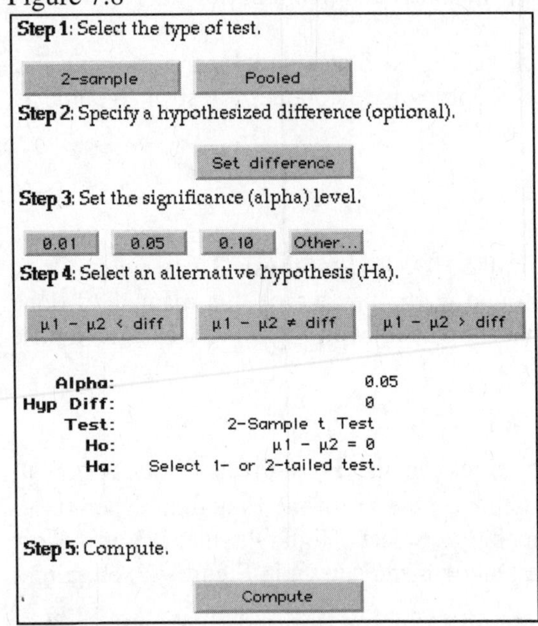

In **Step 1**, we specify how we want DDXL to estimate the variances of the two populations. Two techniques, discussed in the text, can be used. We click on the **2-sample** option as the individual sample variances should be used to estimate their corresponding population variances in this large sample comparison of means. The **Pooled** option would be appropriate when working with small samples in which the assumption of equal population variances is appropriate. **Step 2** asks us to specify the difference in the means that we wish to test. We click on the **Set difference** box to open a menu that requires us to enter a hypothesized difference between the two population means (shown in Figure 7.9). For most comparisons of means, we enter the value **0** in this box and click **OK**. **Step 3** asks us to specify the alpha level that we desire to use for the test. Choices of **0.01**, **0.05**, or **0.10** can be selected or any other value of alpha can be entered by clicking on the **Other…** box and entering the desired value. For this problem, we click on the **0.05** box. **Step 4** requires the user to specify the direction of the test to be computed. The three alternative hypothesis choices $(<, >, and \neq)$ can be selected by clicking on the appropriate box. In this problem, we select a lower tail test by clicking on the $\mu 1 - \mu 2 < diff$ box. Finally, in **Step 5**, we click on **Compute** to create the output shown in Figure 7.10.

Figure 7.9

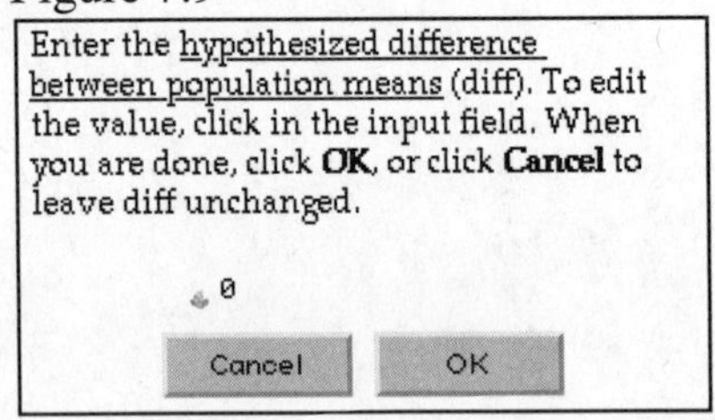

Figure 7.10

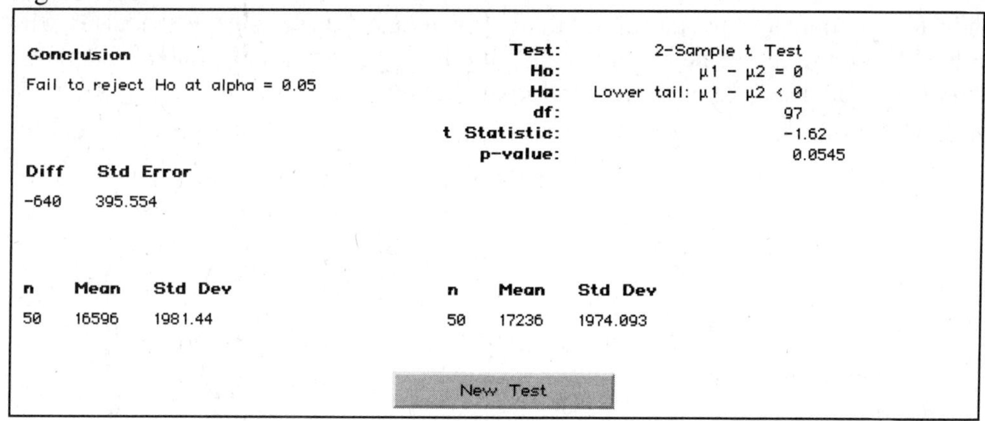

We see from the printout that the test statistic is -1.62 and the computed p-value is .0545. We note that the p-value is slightly different from the one calculated in the text (p = .0526). This is due to DDXL always using t-values instead of using the z-values for large samples as is done in the text. This difference will cause slight discrepancies in the interval endpoints.

7.3 Comparing Two Means – Dependent Sampling

In this section, we will look at how DDXL analyzes data collecting using two data collected from dependent samples. To use this tool within DDXL, **open** a new workbook and highlight the data that is to be compared. Click on the **DDXL Add-In** menu. Click on **Confidence Intervals** or **Hypothesis Tests** option to access the desired menu. Click on the ▼ in the Function Type box to access the different procedures that are available for use. The option that would be selected for the independent comparison of means would be the **Paired t Test** option shown in Figure 7.11.

Figure 7.11

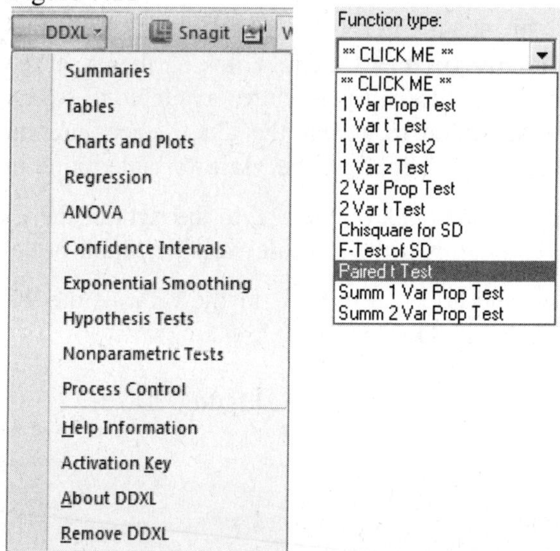

We illustrate how to use these procedures in the following example.

Exercise 7.3: We use Example 7.5 found in the *Statistics for Business and Economics* text.

An experiment is conducted to compare the starting salaries of male and female college graduates who find jobs. Pairs are formed by choosing a male and a female with the same major and similar grade point averages (GPAs). Suppose a random sample of 10 pairs is formed in this manner and the starting salary of each person is recorded. The results are shown in Table 7.2. Compare the mean starting salary, μ_1, for males, to the mean starting salary, μ_2, for females using a 95% confidence interval. Interpret the results.

Table 7.2

PAIR	MALE	FEMALE
1	29300	28800
2	41500	41600
3	40400	39800
4	38500	38500
5	43500	42600
6	37800	38000
7	69500	69200
8	41200	40100
9	38400	38200
10	59200	58500

Solution:

We solve Exercise 7.3 utilizing the **Paired t Test** presented in the DDXL **Confidence Intervals** menu. **Open** the Data File **GRADPAIRS** by following the directions found in the preface of this manual. If done correctly, the data should appear in a workbook similar to that shown below in Figure 7.12. Use the mouse to select the data shown in the workbook.

Figure 7.12

	A	B	C
1	PAIR	MALE	FEMALE
2	1	29300	28800
3	2	41500	41600
4	3	40400	39800
5	4	38500	38500
6	5	43500	42600
7	6	37800	38000
8	7	69500	69200
9	8	41200	40100
10	9	38400	38200
11	10	59200	58500

Click on the **DDXL Add-In** menu. Click on the **Confidence Intervals** option to access the **Confidence Intervals** menu. Click on the ▾ in the Function Type box to access the different procedures available to select. Highlight the **Paired t Test** option to access the Confidence Interval Dialog menu shown in Figure 7.13. Highlight the **Male** variable found in the **Names and Columns Box** and click on the ◀ to the right of the 1st **Quantitative Variable** box. Highlight the **Female** variable found in the **Names and Columns Box** and click on the ◀ to the right of the 2nd **Quantitative Variable** box. Click **OK**.

DDXL now asks you to specify the confidence interval that you wish to create (see Figure 7.14).

Figure 7.13

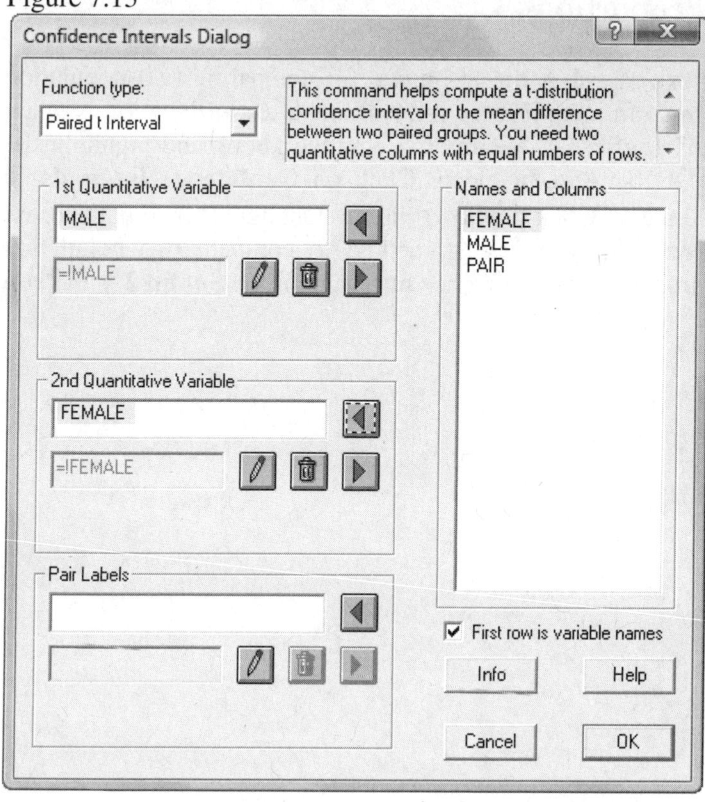

Figure 7.14

We specify the reliability level that we desire to use for the confidence interval. Choices of **90%, 95%,** or **99%** can be selected. For this problem, we click on the **95%** box. We next click on **Compute Interval** to create the output shown in Figure 7.15.

Figure 7.15

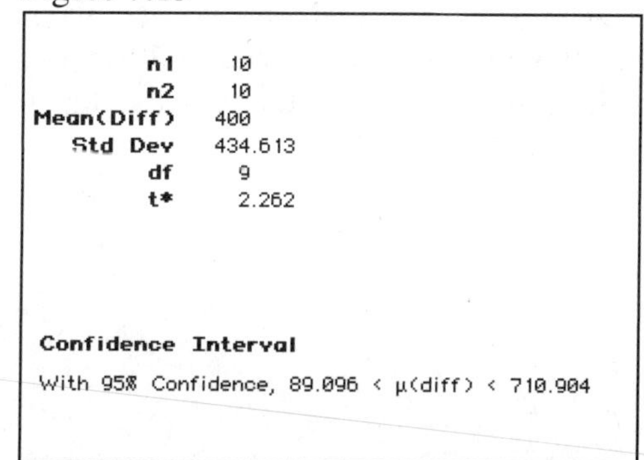

We see from the printout that the computed confidence interval is (89.096, 710.904). We compare these numbers to the ones shown in the text

7.4 Tests For Differences in Two Proportions

The *Statistics for Business and Economics* text describes the technique for comparing two population proportions. DDXL allows the user to perform both a confidence interval and a test of hypothesis when comparing two proportions. To use these tools within DDXL, **open** a new workbook and highlight the data that is to be compared. Click on the **DDXL Add-In** menu. Click on **Confidence Intervals** or **Hypothesis Tests** option to access the desired menu. Click on the ▾ in the Function Type box to access the different procedures that are available for use. The easiest method to compare two population proportions is to use the **Summ 2 Var Prop Interval** (for confidence intervals) or the **Summ 2 Var Prop Test** (for tests of hypothesis).

Figure 7.16

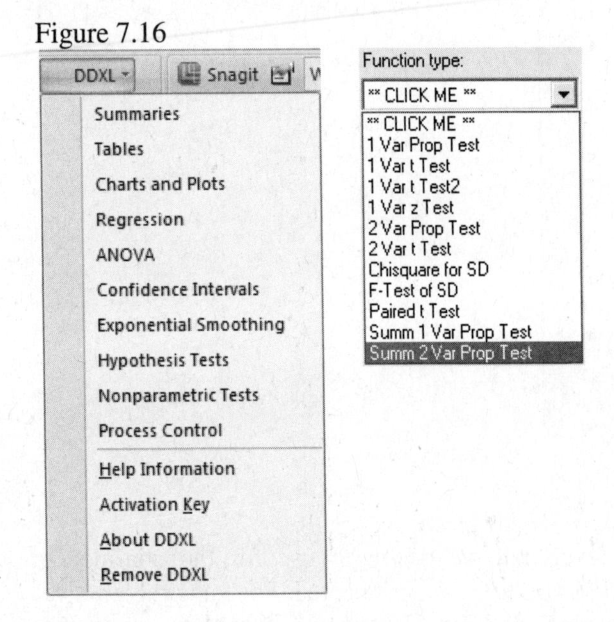

We illustrate how to use these procedures in the following example.

Exercise 7.4: We use Example 7.6 found in the *Statistics for Business and Economics* text.

A consumer advocacy group wants to determine whether there is a difference between the proportions of the two leading automobile models that need major repairs (more than $500) within two years of their purchase. A sample of 400 two-year owners of model 1 is contacted, and a sample of 500 two-year owners of model 2 is contacted. The numbers x_1 and x_2 of owners who report that their cars needed major repairs within the first two years are 53 and 78, respectively. Test the null hypothesis that no difference exists between the proportions in populations 1 and 2 needing major repairs against the alternative that a difference does exist. Use $\alpha = .10$.

Solution:

We solve Exercise 7.4 utilizing the **Summ 2 Var Prop Test** presented in the DDXL **Hypothesis Tests** menu. Open or create the data set **Proportions** shown below in Figure 7.17 and highlight the data. Click on the **DDXL Add-In** menu. Click on the **Hypothesis Tests** option to access the **Hypothesis Tests** menu. Click on the ▾ in the Function Type box to access the different procedures available to select. Highlight the **Summ 2 Prop Test** option to access the Hypothesis Tests Dialog menu shown in Figure 7.18. This menu requires the user to enter the appropriate samples sizes and number of successes in the corresponding spaces in the menu. Highlight the variable names and move them to the appropriate locations using the ◀ icons. The information is entered into the menu and shown below in Figure 7.18. Click OK. DDXL now asks you to specify the confidence interval that you wish to create (see Figure 7.19).

Figure 7.17

A	B	C	D
Sample 1	Success 1	Sample 2	Success 2
400	53	500	78

Figure 7.18

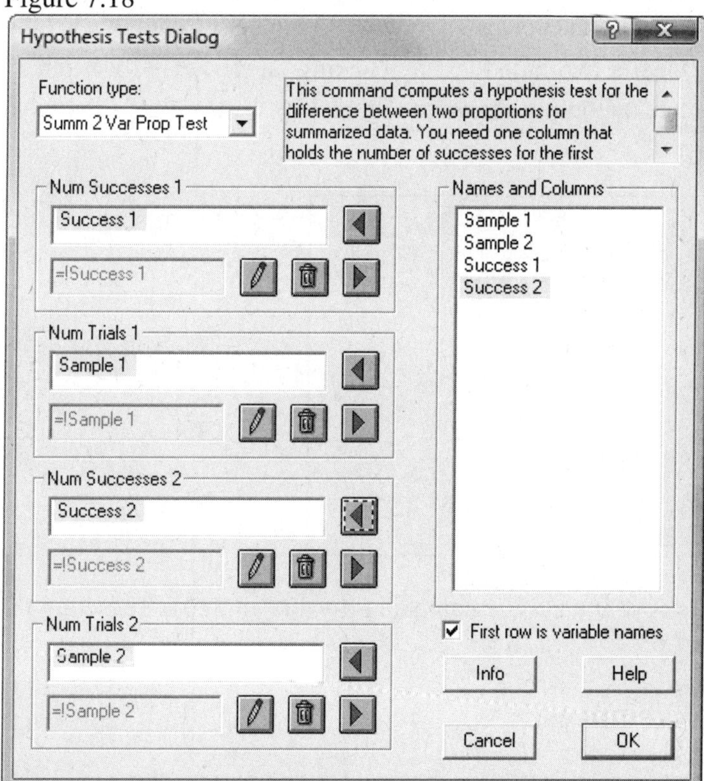

Figure 7.19

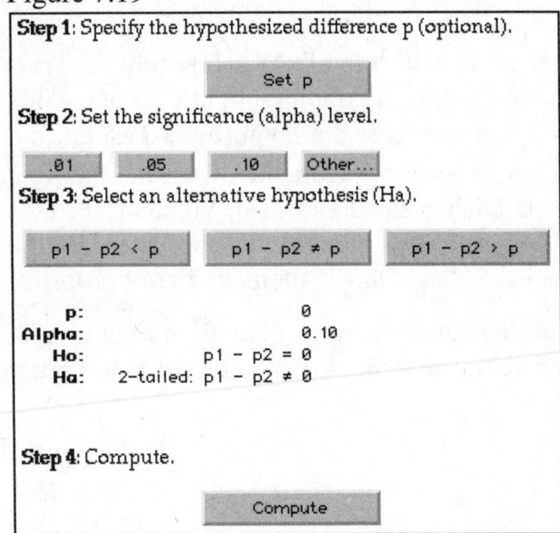

In **Step 1**, we specify the difference in the proportions that we wish to test. We click on the **Set p** box to open a menu that requires us to enter a hypothesized difference between the two population means (shown in Figure 7.20). For most comparisons of proportions, we enter the value **0** in this box and click **OK**.

Figure 7.20

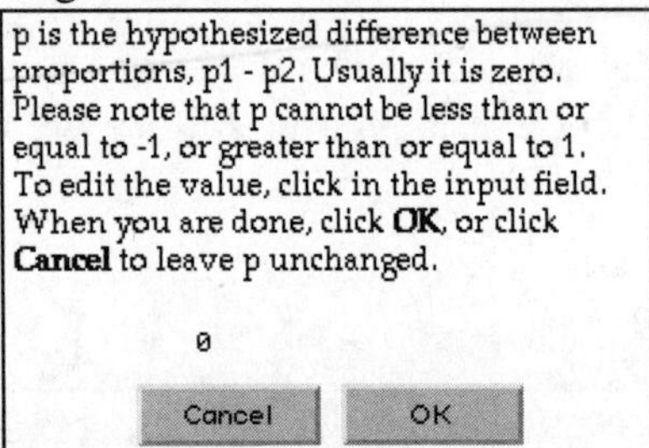

Step 2 asks us to specify the alpha level that we desire to use for the test. Choices of **0.01**, **0.05**, or **0.10** can be selected. Other values of alpha can be selected by clicking on the **Other...** box and entering the value there. For this problem, we click on the **0.10** box. **Step 3** requires the user to specify the direction of the test to be computed. The three alternative hypothesis choices $(<, >, and \neq)$ can be selected by clicking on the appropriate box. In this problem, we select a two-tailed test by clicking on the $p1 - p2 \neq p$ box. Finally, in **Step 4**, we click on **Compute** to create the output shown in Figure 7.21.

Figure 7.21

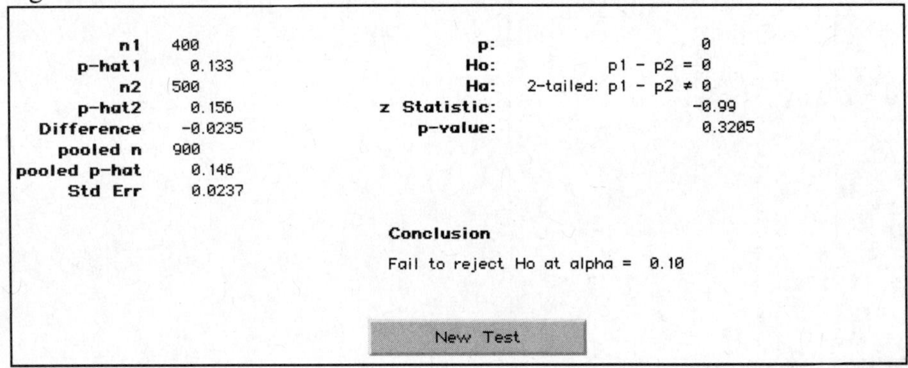

We see from the printout that the test statistic is -0.99 and the computed p-value is .3205. These are the values that are shown in the text for Example 7.6.

7.5 Tests For Differences in Two Variances

The *Statistics for Business and Economics* text describes the technique for comparing two population variances. DDXL allows the user to perform a test of hypothesis when comparing two standard deviations.

To use these tools within DDXL, **open** a new workbook and highlight the data that is to be compared. Click on the **DDXL Add-In** menu. Click on **Confidence Intervals** or **Hypothesis Tests** option to access the desired menu. Click on the ▾ in the Function Type box to access the different procedures that are available for use. The easiest method to compare two population proportions is to use the **F Conf Ints of SD** (for confidence intervals) or the **F Test of SD** (for tests of hypothesis).

Figure 7.22

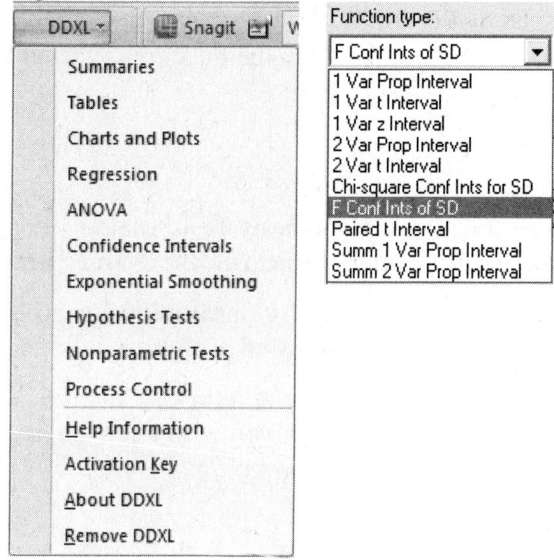

We illustrate how to use these procedures in the following example.

Exercise 7.5: We use Example 7.10 found in the *Statistics for Business and Economics* text.

A manufacturer of paper products wants to compare the variation in daily production levels at two paper mills. Independent random samples of days are selected from each mill and the production levels (in units) recorded. The following summary information was obtained:

	Mill 1	Mill 2
n	13	18
Mean	26.31	19.89
Std. Dev.	8.36	4.85

Do the data provide sufficient evidence to indicate a difference in the variability of production levels at the two paper mills? (Use $\alpha = .10$).

Solution:

We solve Exercise 7.5 utilizing the **F Test of SD** presented in the DDXL **Hypothesis Tests** menu. Open the data set **Papermills** by following the directions found in the preface of this manual. The data is shown below in Figure 7.23. Highlight the data using the mouse. Click on the **DDXL Add-In** menu. Click on the **Hypothesis Tests** option to access the **Hypothesis Tests** menu. Click on the ▾ in the Function Type box to access the different procedures available to select.

Figure 7.23

	A	B
1	Mill 1	Mill 2
2	34	31
3	18	13
4	28	27
5	21	19
6	32	22
7	40	18
8	22	23
9	23	22
10	22	21
11	29	13

Highlight the **F Test of SD** option to access the Hypothesis Tests Dialog menu shown in Figure 7.24. Highlight the **Mill 1** variable found in the **Names and Columns Box** and click on the ◀ to the right of the **1st Quantitative Variable** box. Highlight the **Mill 2** variable found in the **Names and Columns Box** and click on the ◀ to the right of the **2nd Quantitative Variable** box. Click **OK**... DDXL now asks you to specify the test of hypothesis that you wish to create (see Figure 7.25).

Figure 7.24

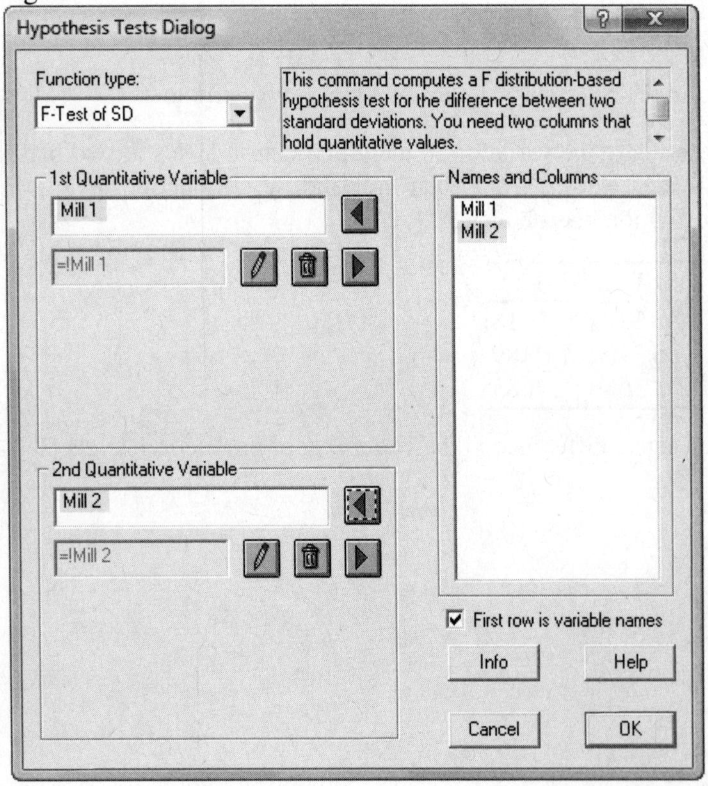

Hypothesis Tests Dialog
Function type:
F-Test of SD
This command computes a F distribution-based hypothesis test for the difference between two standard deviations. You need two columns that hold quantitative values.
1st Quantitative Variable
Mill 1
=!Mill 1
2nd Quantitative Variable
Mill 2
=!Mill 2
Names and Columns
Mill 1
Mill 2
First row is variable names
Info
Help
Cancel
OK

Figure 7.25

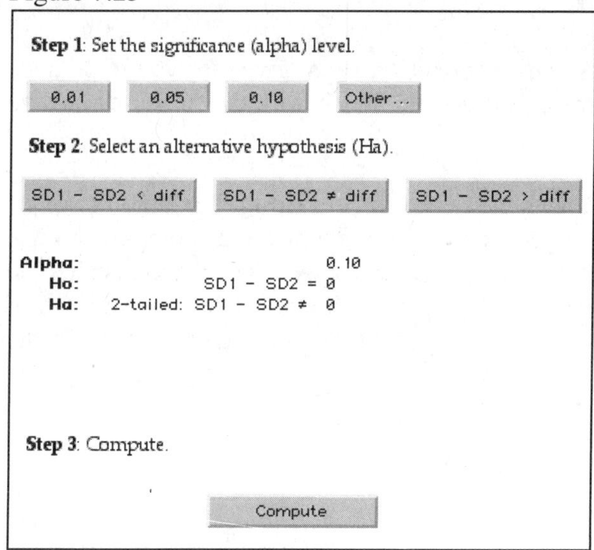

In **Step1**, we specify the alpha level that we desire to use for the test. Choices of **0.01**, **0.05**, or **0.10** can be selected or any other value of alpha can be entered by clicking on the **Other...** box and entering the desired value. For this problem, we click on the **0.10** box. **Step 2** requires the user to specify the direction of the test to be computed. The three alternative hypothesis choices $(<, >, and \neq)$ can be selected by clicking on the appropriate box. In this problem, we select a two-tailed test by clicking on the **SD1-SD2≠ diff**. Finally, in **Step 3**, we click on **Compute** to create the output shown in Figure 7.26.

Figure 7.26

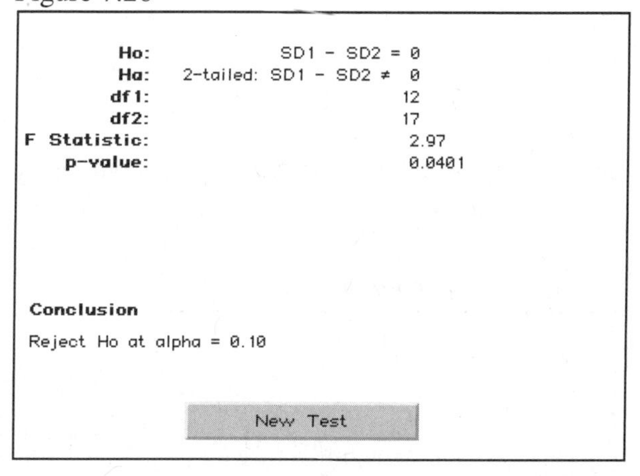

We see from the printout that the test statistic is 2.97 and the computed p-value is .0401. These are the values that are shown in the text for Example 7.10.

Technology Lab

The following exercises from the *Statistics for Business and Economics* text are given for you to practice the procedures covered in the text that are available within DDXL. Included with the exercises are the DDXL outputs that were generated to solve the problems.

7.96 A manufacturer of automobile shock absorbers was interested in comparing the durability of its shocks with that of the shocks produced by its biggest competitor. To make the comparison, one of the manufacturer's and one of the competitor's shocks were randomly selected and installed on the rear wheels of each of six cars. After the cars had been driven 20,000 miles, the strength of each test shock was measured, coded, and recorded. Results of the examination are shown in the table.

Car Number	Manufacturer	Competitor
1	8.8	8.4
2	10.5	10.1
3	12.5	12
4	9.7	9.3
5	9.6	9
6	13.2	13

e. Construct a 95% confidence interval for μ_d.

f. Suppose the data are based on independent random samples. Construct a 95% confidence interval for $\mu_1 - \mu_2$.

DDXL Output 7.94 e

n1	6
n2	6
Mean(Diff)	0.417
Std Dev	0.133
df	5
t*	2.571

Confidence Interval

With 95% Confidence, 0.277 < μ(diff) < 0.556

DDXL Output 7.94 f

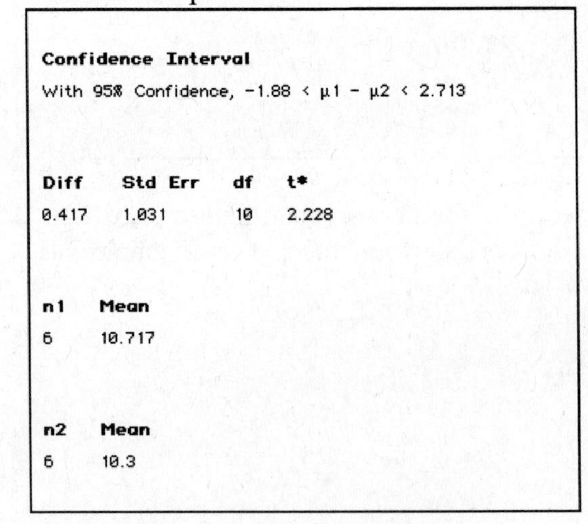

Confidence Interval

With 95% Confidence, -1.88 < μ1 - μ2 < 2.713

Diff	Std Err	df	t*
0.417	1.031	10	2.228

n1	Mean
6	10.717

n2	Mean
6	10.3

7.101. *Industrial Marketing Management* (Vol. 25, 1996) published a study that examined the demographics, decision-making roles, and time demands of product managers. Independent samples of $n_1 = 93$ consumer/commercial product managers and $n_2 = 212$ industrial product managers took part in the study. In the consumer/commercial group, 40% of the product managers are 40 years of age or older; in the industrial group, 54% are 40 or more years old. Make an inference about the difference between the true proportions of consumer/commercial and industrial product managers who are at least 40 years old. Justify your choice of method (confidence interval or hypothesis test) and α level. Do industrial product managers tend to be older than consumer/commercial product managers?

DDXL Output

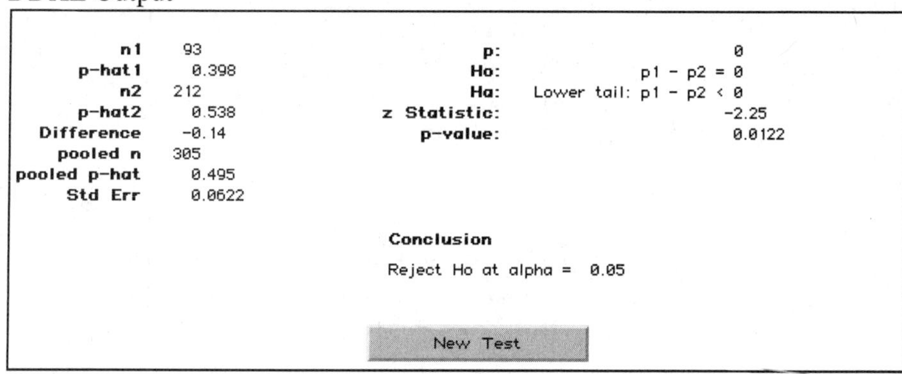

7.112 When new instruments are developed to perform chemical analyses of products (food, medicine, etc.), they are usually evaluated with respect to two criteria: accuracy and precision. *Accuracy* refers to the ability of the instrument to identify correctly the nature and amounts of a product's components. *Precision* refers to the consistency with which the instrument will identify the components of the same material. Thus, a large variability in the identification of a single batch of a product indicates a lack of precision. Suppose a pharmaceutical firm is considering two brands of an instrument designed to identify the components of certain drugs. As part of a comparison of precision, 10 test-tube samples of a well mixed batch of a drug are selected and then 5 are analyzed by instrument A and 5 by instrument B. The data shown below are the percentages of the primary component of the drug given by the instruments. Do these data provide evidence of a difference in the precision of the machines? Use $\alpha = .10$.

Instrument A	Instrument B
43	46
48	49
37	43
52	41
45	48

DDXL output

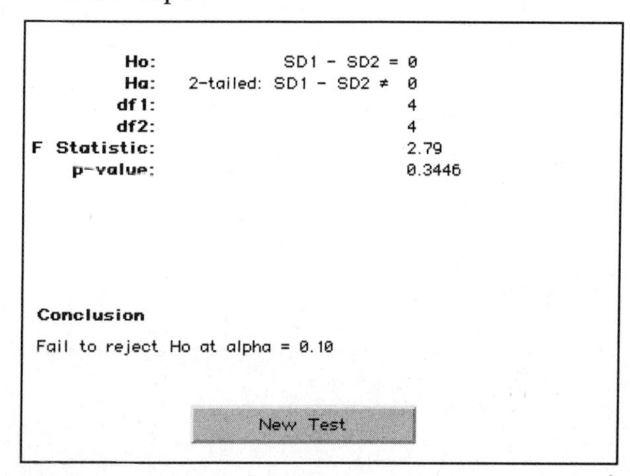

7.107 Marketing strategists would like to predict response to new products and their accompanying promotional schemes. Consequently, studies that examine the differences between buyers and nonbuyers of a product are of interest. A study utilized independent random sample of size 20 and yielded the data shown in the table on the age of the householder primarily responsible for buying toothpaste.

a. Do the data present sufficient evidence to conclude there is a difference in the mean age of purchasers and nonpurchasers? Use $\alpha = .10$.

d. Calculate and interpret a 90% confidence interval for the difference between the mean age of purchasers and nonpurchasers.

Purchasers						Nonpurchasers					
34	35	23	44	52	46	28	22	44	33	55	63
28	48	28	34	33	52	45	31	60	54	53	58
41	32	34	49	50	45	52	52	66	35	25	48
29	59					59	61				

DDXL Output 7.107 a

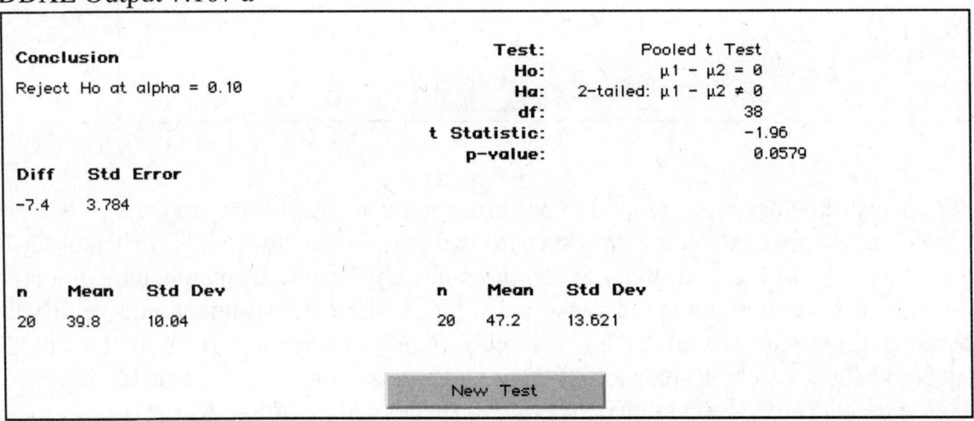

Conclusion

Reject Ho at alpha = 0.10

Test:	Pooled t Test
Ho:	μ1 - μ2 = 0
Ha:	2-tailed: μ1 - μ2 ≠ 0
df:	38
t Statistic:	-1.96
p-value:	0.0579

Diff	Std Error
-7.4	3.784

n	Mean	Std Dev
20	39.8	10.04

n	Mean	Std Dev
20	47.2	13.621

New Test

DDXL Output 7.107 d

Confidence Interval

With 90% Confidence, $-13.779 < \mu 1 - \mu 2 < -1.021$

Diff	Std Err	df	t*
-7.4	3.784	38	1.686

n1	Mean
20	39.8

n2	Mean
20	47.2

Chapter 8
Design of Experiments and Analysis of Variance

8.1 Introduction

Chapter 8 introduces the topics of design of experiments and analysis of variance (ANOVA) to the reader. The concept of the designed experiment is explained and the completely randomized and factorial designs are covered in the text. The goal of analysis of variance is to identify factors that contribute information to the response variable of interest. The combination of levels of the various factors are called treatments and the analysis of variance procedures discussed in the text attempt to detect differences in the mean response variable for the various treatments. Once detected, the text presents several methods of comparing the multiple means of the experiment.

DDXL offers both a one-way a two-way study but neither works as well as the procedures available within Excel. Excel offers two analysis of variance procedures that can be used for the completely randomized and factorial designs. These data analysis tools are very easy to implement. Excel does not, however, offer a follow-up tool to compare treatment means that have been determined to differ. The Excel user must take the summary results from the two analyses and calculate the multiple comparison procedures by hand.

We will use the chapter examples that are given in the text to illustrate the model building and testing methods discussed above. The following examples from *Statistics for Business and Economics* are solved with Microsoft Excel® in this chapter.

Excel Companion Exercise	Page	Statistics for Business and Economics Example	Excel File Name
8.1	132	Example 8.4	GolfCRD
8.2	134	Example 8.10	GolfFAC

8.2 The Completely Randomized Design

The goal of analysis of variance is to compare the mean responses of the various treatments in an experimental design, where the treatments are the combinations of the levels of all the factors involved in the design. The simplest of all experimental designs involves using a single factor to compare values of a response variable. Since there is only one factor in the design, the various levels of the factor are the treatments in the design. The goal is to compare the means of the response variable for those treatments. This experimental design is the completely randomized design and can be analyzed in Excel using the Anova: Single Factor data analysis tool. We illustrate with the following example.

Exercise 8.1: We use Example 8.4 found in the *Statistics for Business and Economics* text.

Suppose the United States Golf Association (USGA) wants to compare the mean distances associated with four different brands of golf balls when struck with a driver. A completely randomized design is employed, with Iron Byron, the USGA's robotic golfer, using a driver to hit a random sample of 10 balls of each brand in a random sequence. The distance is recorded for each hit, and the results are shown in Table 15.1, organized by brand.

a. Set up the test to compare the mean distances for the four brands. Use $\alpha = .10$.
b. Use Excel to obtain the test statistic and p-value. Interpret the results.

Table 8.1

Brand A	Brand B	Brand C	Brand D
251.2	263.2	269.7	251.6
245.1	262.9	263.2	248.6
248.0	265.0	277.5	249.4
251.1	254.5	267.4	242.0
260.5	264.3	270.5	246.5
250.0	257.0	265.5	251.3
253.9	262.8	270.7	261.8
244.6	264.4	272.9	249.0
254.6	260.6	275.6	247.1
248.8	255.9	266.5	245.9

Solution:

Open the Data File **GolfCRD** by following the directions found in the preface of this manual. We click on the **Data** tab and then click on the **Analysis Icon** found in the **Analysis Group**. It should open the **Data Analysis** menu shown below. We highlight the **Anova: Single Factor** option and click **OK**.

Figure 8.1

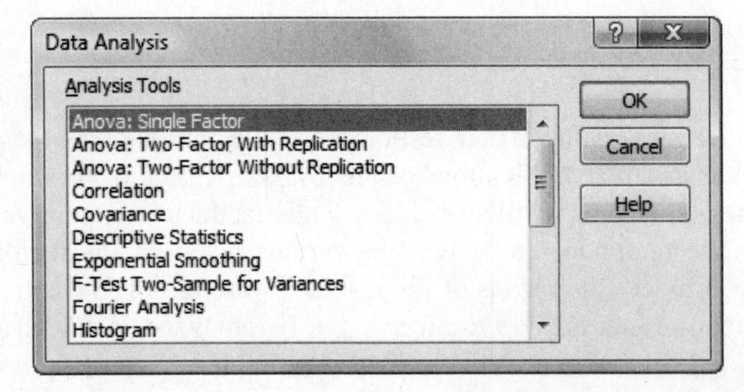

Either type or click the rows and columns where the input data is located and enter this information into the **Input Range** area of the Anova: Single Factor menu (see Figure 8.2). Select the manner in which the data is grouped (**Columns** or **Rows**) and give a level of significance in the **Alpha** cell of the menu (e.g., .10). Specify the location of the computed output by selecting either the **Output Range**, New Worksheet Ply, or New Workbook option, and entering the corresponding **cell** or name. Click **OK**.

Figure 8.2

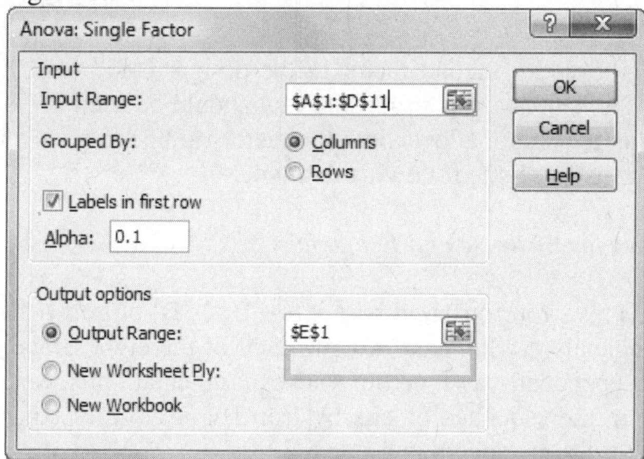

The ANOVA printout generated for the completely randomized design has two main components (see Table 8.2). The first component is a statistical summary of the various treatments in the analysis. For each of the four brands of balls, Excel gives some summary information concerning the distances achieved by each. This information will be more useful after studying the multiple comparison of means material in Section 8.3 of the text.

Table 8.2

Anova: Single Factor

SUMMARY

Groups	Count	Sum	Average	Variance
Brand A	10	2507.8	250.78	22.42177778
Brand B	10	2610.6	261.06	14.94711111
Brand C	10	2699.5	269.95	20.25833333
Brand D	10	2493.2	249.32	27.07288889

ANOVA

Source of Variation	SS	df	MS	F	P-value	F crit
Between Groups	2794.389	3	931.462917	43.98874592	3.97311E-12	2.242607877
Within Groups	762.301	36	21.1750278			
Total	3556.69	39				

The second component is called the analysis of variance table and is where the pertinent testing information will be found. To test whether the mean distances of the four means differ, we use the test statistic and p-value found in the Brand row of the printout (labeled on the Excel printout as the Between Groups row). We see that the test statistic is $F = 43.9887$ and the p-value is $p \approx 0$. Compare these values to the values found in the Excel printout found in the text. We refer you to the text for further information regarding the interpretation of these values.

8.3 The Factorial Design

The next step in the experimental design process is to add a second factor to the design. One possible design that results is the factorial design. In Excel, the data analysis procedure that should be used is the Anova: Two-Factor With Replication procedure. This procedure allows both the factors to be analyzed as well as the interaction between them. We illustrate its use with the following example.

Exercise 8.2: We use Example 8.10 in the *Statistics for Business and Economics* text.

Suppose the United States Golf Association (USGA) tests four different brands (A, B, C, D) of golf balls and two different clubs (driver, five-iron) in a completely randomized design. Each of the eight Brand-Club combinations (treatments) is randomly and independently assigned to four experimental units, each experimental unit consisting of a specific position in the sequence of hits by Iron Byron. The distance response is recorded for each of the 32 hits, and the results are shown in Table 8.3.

a. Use Excel to partition the Total Sum of Squares into the components necessary to analyze this 4x2 factorial experiment.
b. Follow the steps for analyzing a two-factor factorial experiment and interpret the results of your analysis. Use $\alpha = .10$ for the tests you conduct.

Table 8.3

		BRAND			
		A	B	C	D
	DRIVER	226.4	238.3	240.5	219.8
	DRIVER	232.6	231.7	246.9	228.7
	DRIVER	234.0	227.7	240.3	232.9
CLUB	DRIVER	220.7	237.2	244.7	237.6
	FIVE-IRON	163.8	184.4	179.0	157.8
	FIVE-IRON	179.4	180.6	168.0	161.8
	FIVE-IRON	168.6	179.5	165.2	162.1
	FIVE-IRON	173.4	186.2	156.5	160.3

Solution:

Open the Data File **GolfFAC** by following the directions found in the preface of this manual. We click on the **Data** tab and then click on the **Analysis Icon** found in the **Analysis Group**. It should open the **Data Analysis** menu shown below in Figure 8.3. We highlight the **Anova: Two-Factor With Replication** option and click **OK**.

Figure 8.3

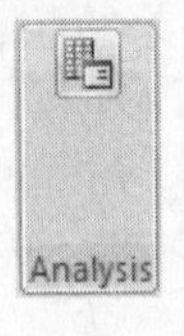

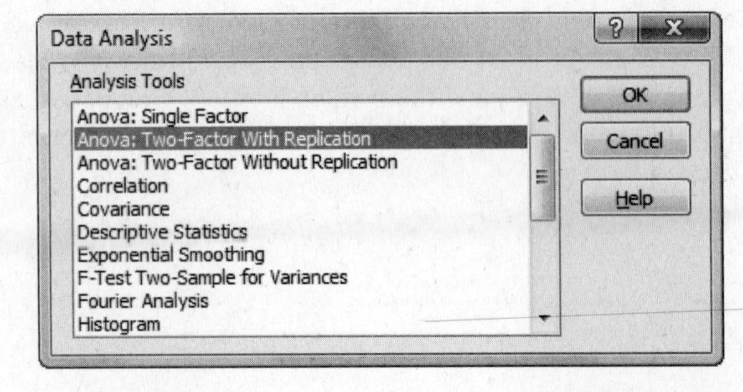

Figure 8.4

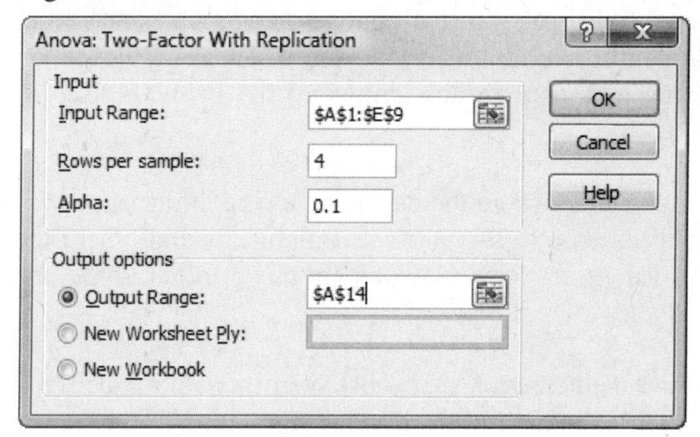

Either type or click the rows and columns where the input data is located and enter this information into the **Input Range** area of the Anova: Two Factor With Replication menu. Include the labels for the columns and rows of the factor when inputting the data range. Note that each Brand of golf ball includes four rows of data for each of the two Clubs tested. Enter the number of rows (e.g., 4) in the **Rows per Sample** area of the menu. Give a level of significance in the **Alpha** cell of the menu (e.g., .10). Specify the location of the computed output by selecting either the **Output Range**, New Worksheet Ply, or New Workbook option, and entering the corresponding **cell** or name. Click **OK**.

Table 8.4

Anova: Two-Factor With Replication

SUMMARY	A	B	C	D	Total
DRIVER					
Count	4	4	4	4	16
Sum	913.7	934.9	972.4	919	3740
Average	228.425	233.725	243.1	229.75	233.75
Variance	37.429167	24.46917	10.53333	57.21667	61.07067
FIVE-IRON					
Count	4	4	4	4	16
Sum	685.2	730.7	668.7	642	2726.6
Average	171.3	182.675	167.175	160.5	170.4125
Variance	44.52	9.929167	86.1225	3.86	98.19183
Total					
Count	8	8	8	8	
Sum	1598.9	1665.6	1641.1	1561	
Average	199.8625	208.2	205.1375	195.125	
Variance	967.48268	759.3429	1688.454	1396.336	

ANOVA

Source of Variation	*SS*	*df*	*MS*	*F*	*P-value*	*F crit*
Sample	32093.111	1	32093.11	936.7516	9.63E-21	2.927116
Columns	800.73625	3	266.9121	7.790779	0.00084	2.32739
Interaction	765.96125	3	255.3204	7.452435	0.001079	2.32739
Within	822.24	24	34.26			
Total	34482.049	31				

The ANOVA printout generated for the factorial design has two main components (see Table 8.4). The first component is a statistical summary of the various treatments in the analysis. For each of the eight Brand-Club treatments, Excel gives some summary information concerning the distances achieved by each. This information can be used when comparing treatment means similar to the methods used in Section 8.3.

The second component is the analysis of variance table and is where the pertinent testing information will be found. This is where the sums of squares are petitioned into the various components and where the interaction and individual factor test statistics and p-values are located. Compare this printout to the one found in the text.

The first test of interest to the USGA is to determine if interaction exists between the Club and Brand factors in the experiment. We use the test statistic ($t = 7.452435$) and the p-value ($p = 0.001079$) found in the interaction row of the analysis of variance table. Refer to the text for more information concerning the interpretation of these values and the follow-up analysis that is necessary for factorial designs.

We note here one drawback associated with the Excel analysis for the completely randomized and factorial designs in the analysis of variance experiments. Excel offers the appropriate analyses for determining when differences exist between the treatment means for each of these two experimental designs, but does not offer any method to determine where the specific differences exists. Both the SAS™ and MINITAB™ software packages offer options for conducting the multiple comparison procedures that enable the user to conduct the appropriate follow-up analysis for both the completely randomized and factorial designs. Consult Section 8.3 and the references at the end of the text for more information on this topic.

Chapter 9
Categorical Data Analysis

9.1 Introduction

Chapter 9 introduces the topic of categorical data analysis to the reader. Both the one-way and two-way analyses are discussed in the text. DDXL allows the user to work with both of these analyses. We illustrate both using the following examples from the text:

Excel Companion Exercise	Page	Statistics for Business and Economics Example	Excel File Name
9.1	139	Example 9.2	Payplan
9.2	142	Example 9.3	Brokerage

9.2 Testing Categorical Probabilities: One-Way Table

The one-way analysis of categorical probabilities allows the user to compare observed counts of the levels of a qualitative variable against some hypothesized probabilities. The user has the flexibility of hypothesizing equal probabilities or individual probabilities, provided that they all sum to the value one. DDXL requires the user to create a data set that contains three columns – one that contains the levels of the variable being tested, a second that contains the observed counts for these levels, and a third that contains the hypothesized proportions (in decimal form) for these levels. We illustrate with the following example.

Exercise 9.1 Use example 9.2 found in the *Statistics for Business and Economics* text.

A large firm has established what it hopes is an objective system of deciding on annual pay increases for its employees. The system is based on a series of evaluation scores determined by the supervisors of each employee. Employees with scores above 80 receive a merit pay increase, those with scores between 50 and 80 receive the standard increase, and those below 50 receive no increase. The firm designed the plan with the objective that, on the average, 25% of its employees would receive merit increases, 65% would receive standard increases, and 10% would receive no increase. After 1 year of operation using the new plan, the distribution of pay increases for a random sample of 600 company employees was as shown in Table 9.1. Test at the $\alpha = .01$ level to determine whether these data indicate that the distribution of pay increases differs significantly from the proportions established by the firm.

Table 9.1

None	Standard	Merit
42	365	193

Solution:

We solve Exercise 9.1 utilizing the **Goodness of Fit** option presented in the DDXL **Tables** menu (see Figure 9.1). **Open** the Data File **Payplan** by following the directions found in the preface of this manual. It should include three columns of data columns – one that contains the three levels (None, Standard and Merit) of the pay plan variable being tested, a second that contains the observed counts (42, 365, and 193) for these levels, and a third that contains the hypothesized proportions (.10, .65, and .25) for these levels. If done correctly, the data should appear in a workbook similar to that shown below in Figure 9.2. Use the mouse to select the data shown in the workbook.

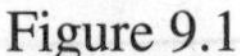

Figure 9.1

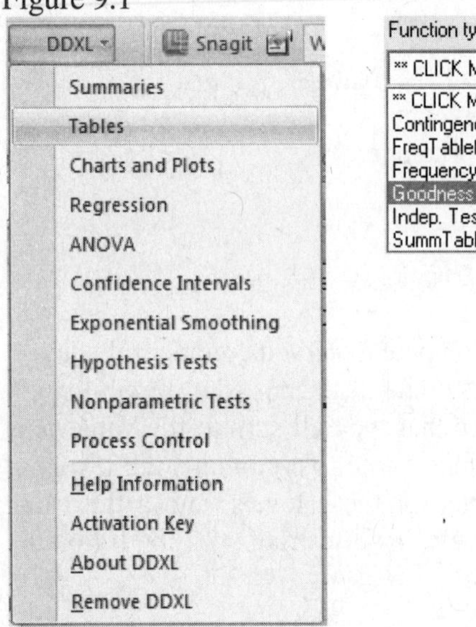

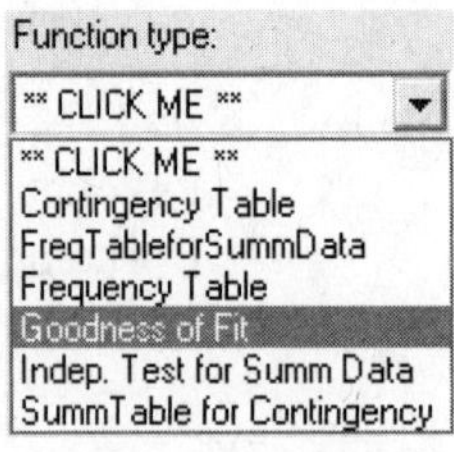

Figure 9.2

A	B	C
Type	Count	Expected
None	42	0.1
Standard	365	0.65
Merit	193	0.25

Click on the **DDXL Add-In** menu. Click on the **Tables** option to access the **Tables** menu. Click on the ▾ in the Function Type box to access the different procedures available to select. Highlight the **Goodness of Fit** option to access the Tables Dialog menu shown in Figure 9.3. Highlight the **Type** variable found in the **Names and Columns Box** and click on the ◀ to the right of the **Category Names** box. Highlight the **Count** variable found in the **Names and Columns Box** and click on the ◀ to the right of the **Observed Counts** box. Highlight the **Expected** variable found in the **Names and Columns Box** and click on the ◀ to the right of the **Test Distribution** box. Click **OK**. DDXL gives us the output shown below in Figure 9.4.

Figure 9.3

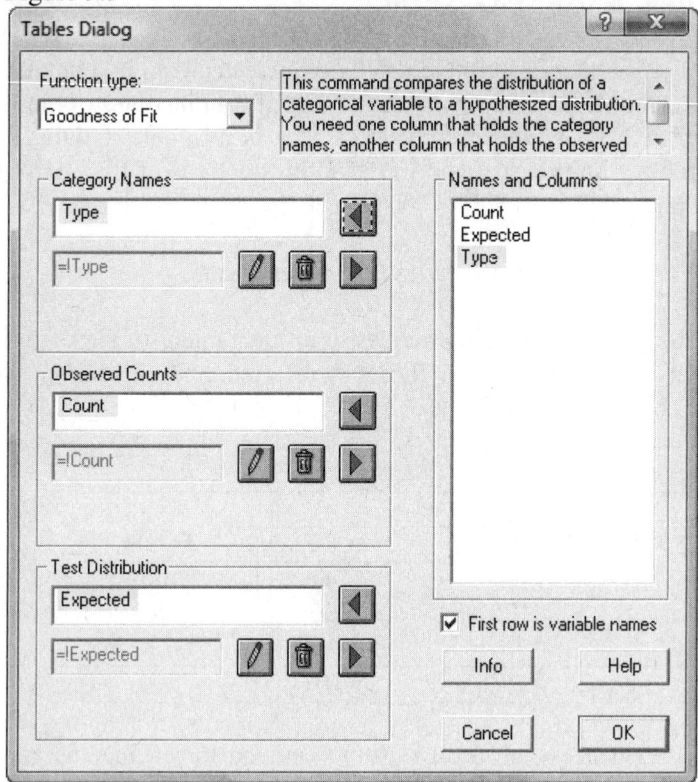

Figure 9.4

Type	Count	Expected	Expected Frequencies
None	42	0.1	60
Standard	365	0.65	390
Merit	193	0.25	150

All Exp. Freqs >= 1?	Assumption Met
At Most 20% of Exp. Freqs < 5?	Assumption Met

chi-square	19.329
p-value	< 0.0001

The DDXL output contains three sections. The first is a table that contains the data that was entered into the Excel spreadsheet. The second part looks at a couple of common assumptions required by the Goodness of Fit test. The third section contains the results of the test. We compare the test statistic of 19.329 and the p-value of $p < .0001$ to the corresponding values listed in the text.

9.3 Testing Categorical Probabilities: Two-Way Table

Chapter 2 introduced the reader to the idea of presenting descriptive results of collected data in a tabular form. In Chapter 9, we now take a look at a technique that allows us to determine if the outcomes of these two variables are dependent upon one another. The two-way analysis of data is available in DDXL through using the **Independent Test of Summary Data** test found in the **Tables** options of the **DDXL** menu. We illustrate using the following exercise.

Exercise 9.2 Use example 9.3 found in the *Statistics for Business and Economics* text.

A large brokerage firm wants to determine whether the service it provides to affluent clients differs from the service it provides to lower-income clients. A sample of 500 clients is selected, and each client is asked to rate his or her broker. The results are shown below in Table 9.2.

Table 9.2

		Clients Income			
		Under $30,000	**$30-60,000**	**Over $60,000**	**Totals**
	Outstanding	48	64	41	153
Broker	**Average**	98	120	50	268
Rating	**Poor**	30	33	16	79
	Totals	176	217	107	500

a. Test to determine whether there is evidence that broker rating and customer income are independent. Use $\alpha = .10$.

Solution:

We solve Exercise 9.2 utilizing the **Indep. Test for Summ Data** option presented in the DDXL **Tables** menu (see Figure 9.6). **Open** the Data File **Brokerage** by following the directions found in the preface of this manual. It should include three columns of data columns – one that contains the three levels (Outstanding, Average, and Poor) of the Broker Rating variable being tested, a second that contains the three levels (Under, Between, and Over) of the Clients Income being tested, and a third that contains the observed counts for these levels. If done correctly, the data should appear in a workbook similar to that shown below in Figure 9.5. Use the mouse to select the data shown in the workbook.

Figure 9.5

	A	B	C
1	**Broker Rating**	**Client's Income**	**Count**
2	Outstanding	Under	48
3	Outstanding	Between	64
4	Outstanding	Over	41
5	Average	Under	98
6	Average	Between	120
7	Average	Over	50
8	Poor	Under	30
9	Poor	Between	33
10	Poor	Over	16

Figure 9.6

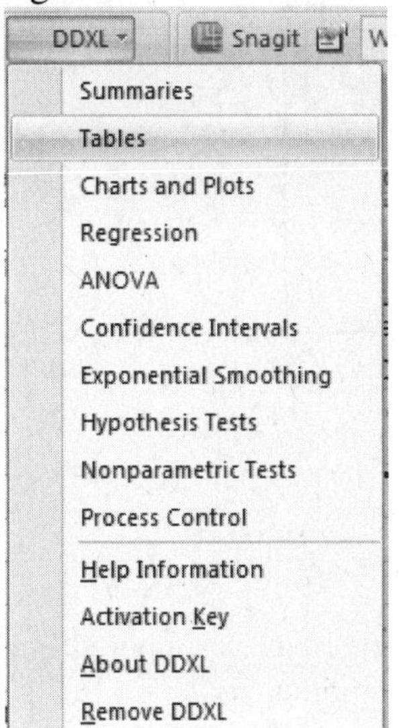

Click on the **DDXL Add-In** menu. Click on the **Tables** option to access the **Tables** menu. Click on the ▾ in the Function Type box to access the different procedures available to select. Highlight the **Indep. Test for Summ Data** option to access the Tables Dialog menu shown in Figure 9.7. Highlight the **Broker Rating** variable found in the **Names and Columns Box** and click on the ◀ to the right of the **Variable One Names** box. Highlight the **Client's Income** variable found in the **Names and Columns Box** and click on the ◀ to the right of the **Variable Two Names** box. Highlight the **Counts** variable found in the **Names and Columns Box** and click on the ◀ to the right of the **Counts** box. Click **OK**. DDXL gives us the output shown below in Figure 9.8.

Figure 9.7

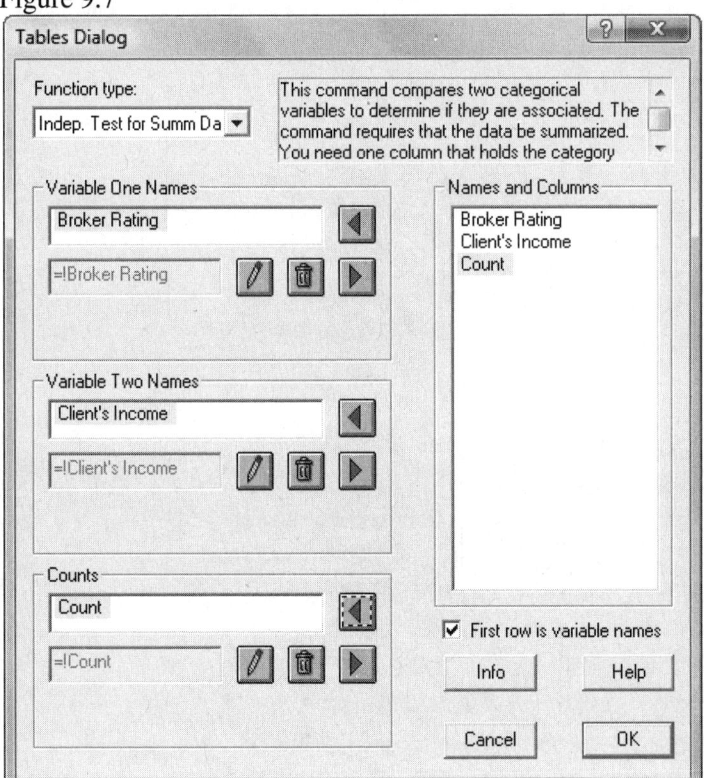

Figure 9.8

Broker Rating	Client's Income	Count	Row Total	Column Total	Expected Frequencies
Outstanding	Under	48	153	176	53.856
Outstanding	Between	64	153	217	66.402
Outstanding	Over	41	153	107	32.742
Average	Under	98	268	176	94.336
Average	Between	120	268	217	116.312
Average	Over	50	268	107	57.352
Poor	Under	30	79	176	27.808
Poor	Between	33	79	217	34.286
Poor	Over	16	79	107	16.906

All Exp. Freqs >= 1?	Assumption Met
At Most 20% of Exp. Freqs < 5?	Assumption Met

chi-square	4.278
p-value	0.3697

The DDXL output contains three sections. The first is a table that contains the data that was entered into the Excel spreadsheet. The second part looks at a couple of common assumptions required by the Goodness of Fit test. The third section contains the results of the test. We compare the test statistic of 4.278 and the p-value of $p < .3697$ to the corresponding values listed in the text.

Technology Lab

The following exercises from the *Statistics for Business and Economics* text is given for you to practice the categorical two-way procedure that is available within DDXL. Included with the exercise is the DDXXL output that was generated to solve the problem.

9.36 *Bon Appetit* magazine polled 200 of its readers concerning which of the four vegetables – brussel sprouts, okra, lima beans, and cauliflower – is their least favorite. The results (adapted from *Adweek*, Feb. 21, 2000) are presented in the table. Let $p_1, p_2, p_3,$ and p_4 represent the proportions of all *Bon Appetit* readers who indicate brussel sprouts, okra, lima beans, and cauliflower, respectively, as their least favorite vegetable.

Brussel Sprouts	Okra	Lima Bean	Cauliflower
46	76	44	34

Conduct a test to determine whether *Bon Appetit* readers have a preference for one of the vegetables as "least favorite."

DDXL Output

Vegetable	Count	Expected	Expected Frequencies
Brussel Sprouts	46	0.25	50
Okra	76	0.25	50
Lima Bean	44	0.25	50
Cauliflower	34	0.25	50

All Exp. Freqs >= 1?	Assumption Met
At Most 20% of Exp. Freqs < 5?	Assumption Met

chi-square	19.68
p-value	0.0002

9.28 An article in *Sociological Methods & Research* (May, 2001) analyzed the data presented in the table. A sample of 262 Kansas pig farmers were classified according to their education level (college or not) and size of their pig farm (number of pigs). Conduct a test to determine whether a pig farmer's education level has an impact on the size of the pig farm. Use $\alpha = .05$ and support your answer with a graph.

Data

	Education Level		
Farm Size	**No College**	**College**	**TOTALS**
<1,000 pigs	42	53	95
1,000-2,000 pigs	27	42	69
2,000-5,000 pigs	22	20	42
>5,000 pigs	27	29	256
TOTALS	118	144	262

DDXL Output

Size	College	Count	Row Total	Column Total	Expected Frequencies
A	No	42	95	118	42.786
A	Yes	53	95	144	52.214
B	No	27	69	118	31.076
B	Yes	42	69	144	37.924
C	No	22	42	118	18.916
C	Yes	20	42	144	23.084
D	No	27	56	118	25.221
D	Yes	29	56	144	30.779

All Exp. Freqs >= 1?	Assumption Met
At Most 20% of Exp. Freqs < 5?	Assumption Met

chi-square	2.142
p-value	0.5434

Chapter 10
Simple Linear Regression

10.1 Introduction

Chapters 10 and 11 in *Statistics for Business and Economics* introduce the topic of regression analysis to the reader. Chapter 10 serves as the introduction of the general concepts of simple linear regression. Simple Linear Regression is how the text introduces the theories and concepts of mathematical modeling to the reader. These topics are then expanded in Chapter 11 of the text.

We will take a similar approach to regression as does the text. We will use Chapter 10 to introduce you to the methods DDXL offers to work with regression analysis. We will see how DDXL can be used to calculate both the correlation and the linear modeling ideas that are presented in the text. We will use the chapter examples that are given in the text to illustrate these methods. The following examples from *Statistics for Business and Economics* are solved with DDXL in this chapter:

Excel Companion Exercise	Page	Statistics for Business and Economics Example	Excel File Name
10.1	147	Example 10.4	Casino
10.2	150	Example 10.5	Adsales
10.3	153	Example 10.6/10.7	Adsales

10.2 The Coefficient of Correlation

Regression analysis is all about the relationship between variables. Chapters 10 and 11 spend time developing the mathematical modeling of one variable using the values of other related variables. The simplest form of this modeling idea is the linear relationship between two variables. This idea, known as correlation, is studied in Chapter 10 of *Statistics for Business and Economics*. We examine how DDXL calculates correlations below.

Exercise 10.1: Use Example 10.4 found in the *Statistics for Business and Economics* text.

Legalized gambling is available on several riverboat casinos operated by a city in Mississippi. The mayor of the city wants to know the correlation between the number of casino employees and the yearly crime rate. The records for the past 10 years are examined and the results listed in Table 10.1 are obtained. Calculate the coefficient of correlation r for the data.

Table 10.1

Year	Number of Employees, x	Crime Rate, y
2000	15	1.35
2001	18	1.63
2002	24	2.33
2003	22	2.41
2004	25	2.63
2005	29	2.93
2006	30	3.41
2007	32	3.26
2008	35	3.63
2009	38	4.15

Solution:

We solve Exercise 10.1 utilizing the **Correlation** function presented in the DDXL **Regression** menu (see Figure 10.1). **Open** the Data File **Casino** by following the directions found in the preface of this manual. If done correctly, the data should appear in a workbook similar to that shown below in Figure 10.2. Use the mouse to select the data shown in the workbook.

Figure 10.1

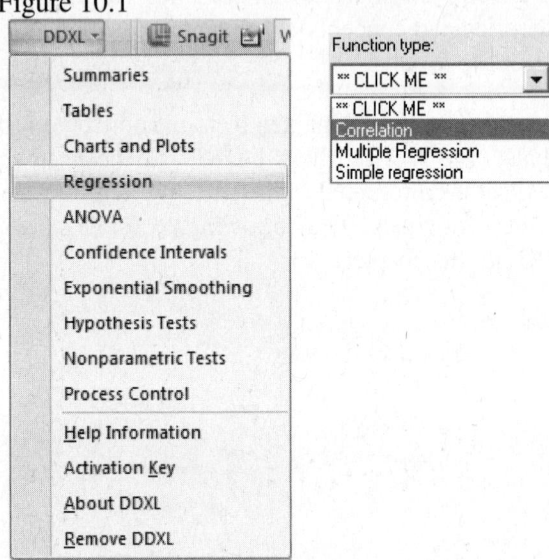

Figure 10.2

	A	B	C
1	Year	Employees	Crime
2	2000	15	1.35
3	2001	18	1.63
4	2002	24	2.33
5	2003	22	2.41
6	2004	25	2.63
7	2005	29	2.93
8	2006	30	3.41
9	2007	32	3.26
10	2008	35	3.63
11	2009	38	4.15

Click on the **DDXL Add-In** menu. Click on the **Regression** option and click on the ▾ in the Function Type box to access the different procedures available to select. Highlight the **Correlation** option to access the Regression Dialog menu shown in Figure 10.3. Highlight the **Employees** variable found in the **Names and Columns Box** and click on the ◀ to the right of the **x-Axis Quantitative Variable** box. Next, we highlight the **Crime** variable found in the **Names and Columns Box** and click on the ◀ to the right of the **y-Axis Quantitative Variable** box. Click **OK**. DDXL gives us the output shown below in Figure 10.4.

Figure 10.3

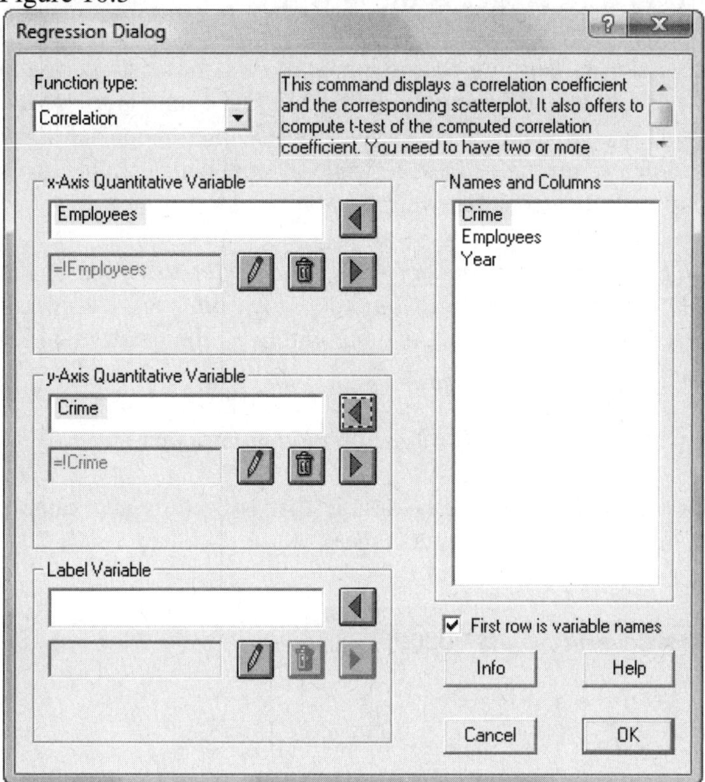

Figure 10.4

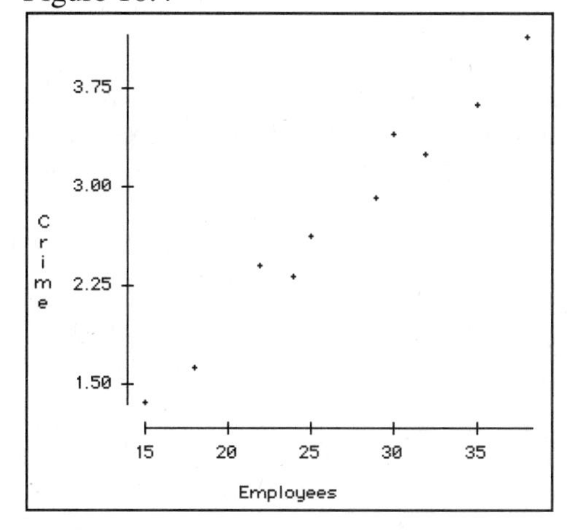

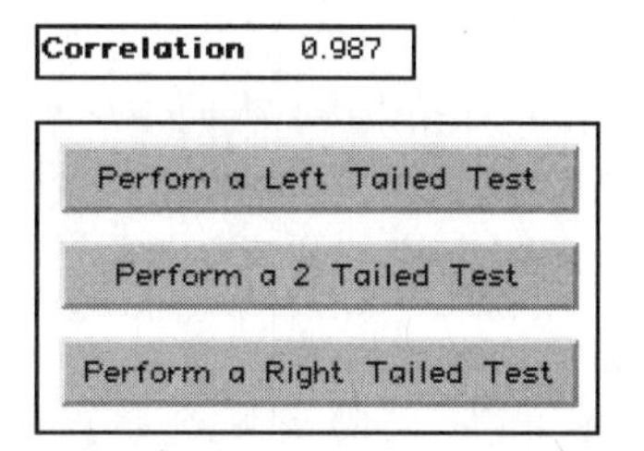

The DDXL output contains several parts. First, a quick plot of the data is drawn to get a picture of how the variables are related to each other. Second, the coefficient of correlation is calculated. For this data, the correlation is calculated to be 0.987. Finally, DDXL offers the user the chance to conduct a correlation t-test. DDXL asks the user to specify the direction of the test to conduct. While no t-test is desired for this problem, the results of the upper tail test are shown in Figure 10.5 for illustration purposes.

Figure 10.5

Ha	Two Tailed
t-statistic	17.39
p-value	< .0001

We next look at how the coefficient of determination and other simple linear regression output are generated within DDXL.

10.3 The Coefficient of Determination and Regression Output

After studying the topic of correlation, the next step in learning regression analysis is understanding the modeling concepts. Our goal in regression is to build a mathematical relationship that attempts to predict the value of one variable, y, with the values of other related variables, the x's. Chapter 10 presents the simplest form of this modeling idea -- simple linear regression. In it, a single independent variable, x, is hypothesized to have a straight-line relationship with the dependent variable, y.

The example that we use from *Statistics for Business and Economics* asks the reader to calculate the coefficient of determination from the data. Our purpose here is to use the data from the example to generate the basic simple linear regression output with Excel. As you will see, the coefficient of determination is one of the components of this output.

Exercise 10.2: We use Example 10.5 found in the *Statistics for Business and Economics* text.

Calculate the coefficient of determination for the advertising-sales example that is used as an example throughout the text. The data are shown below in Table 10.2 for convenience.

Table 10.2

Advertising Expenditures, x (100s)	Sales Revenues, y ($1,000s)
1	1
2	1
3	2
4	2
5	4

Solution:

We solve Exercise 10.2 utilizing the **Simple Regression** function presented in the DDXL **Regression** menu (see Figure 10.6). **Open** the Data File **Adsales** by following the directions found in the preface of this manual. If done correctly, the data should appear in a workbook similar to that shown below in Figure 10.7. Use the mouse to select the data shown in the workbook.

Figure 10.6

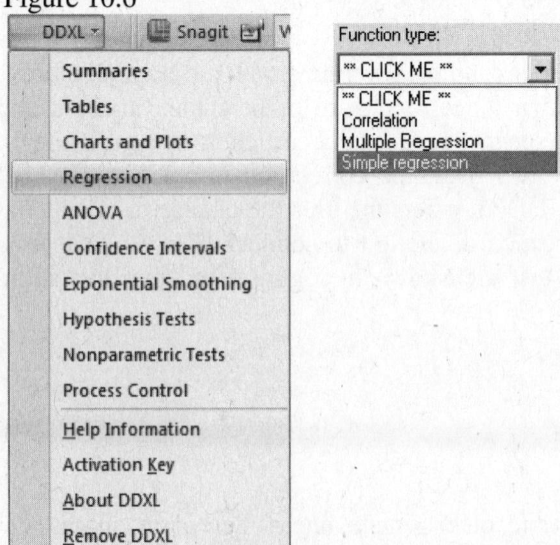

Figure 10.7

	A	B
1	ADVERTISING	SALES
2	1	1
3	2	1
4	3	2
5	4	2
6	5	4

Click on the **DDXL Add-In** menu. Click on the **Regression** option and click on the in the Function Type box to access the different procedures available to select. Highlight the **Simple Regression** option to access the Regression Dialog menu shown in Figure 10.8. Highlight the **SALES** variable found in the **Names and Columns Box** and click on the to the right of the **Response Variable** box. Next, we highlight the **ADVERTISING** variable found in the **Names and Columns Box** and click on the to the right of the **Explanatory Variable** box. Click **OK**. DDXL gives us the output shown below in Figure 10.9.

Figure 10.8

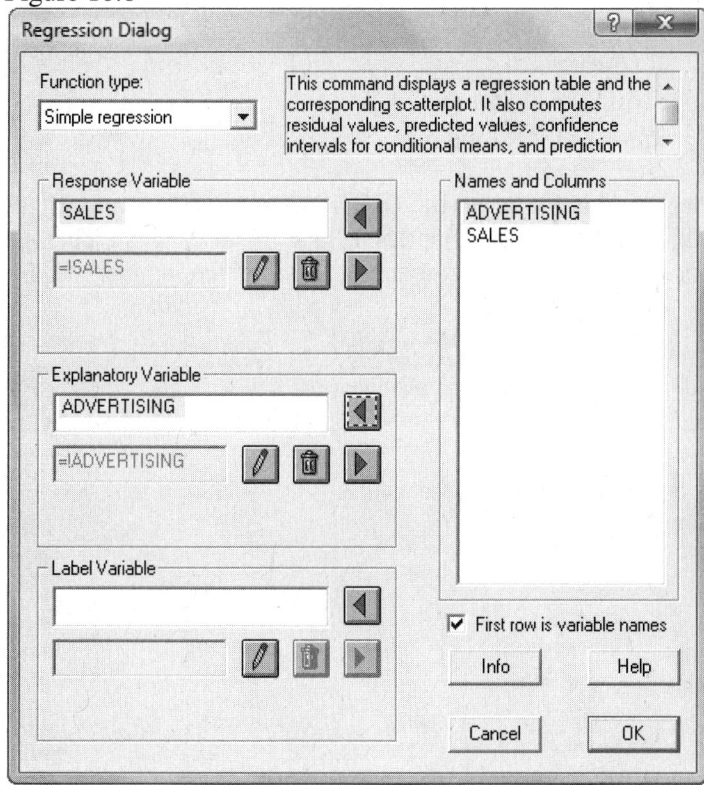

Figure 10.9

Dependent variable is: SALES
No Selector
R squared = 81.7% R squared (adjusted) = 75.6%
s = 0.6055 with 5 - 2 = 3 degrees of freedom

Source	Sum of Squares	df	Mean Square	F-ratio
Regression	4.9	1	4.9	13.4
Residual	1.1	3	0.366667	

Variable	Coefficient	s.e. of Coeff	t-ratio	prob
Constant	-0.1	0.6351	-0.157	0.8849
ADVERTIS...	0.7	0.1915	3.66	0.0354

DDXL provides several different types of output as part of its standard Simple Regression results. Figure 10.9 shows the basic simple linear regression output that is provided by most statistical software. In addition to the coefficient of determination desired in this exercise, DDXL provides many other useful regression results. We point out the most important features of the printout generated by DDXL below. Compare these values to those shown in the *Statistics for Business and Economics* text.

DDXL Printout Values	**Description of Values**
R Square = 81.7%	Coefficient of determination
s = 0.6055	Standard Deviation
Constant Coefficient = -0.1	Estimate of β_o
ADVERTIS_ Coefficient = 0.7	Estimate of β_1
t-ratio for ADVERTIS_= 3.66	Test Statistic for testing β_1
P-value for ADVERTIS_= 0.0354	P-value for testing β_1

We find the value of the coefficient of determination is found on the DDXL to be R^2 =81.7%. Many of the other values presented in Figure 10.9 will be discussed in Chapters 10 and 11 in the text. The last topic that we cover in Chapter 10 is the topic of generating confidence and prediction intervals for specified values of X.

10.4 Estimating and Predicting with a Simple Linear Model

The final step in the simple linear regression analysis is to use the model to estimate and predict values of the dependent variable, y, for specified settings of the independent variable, x. We illustrate this procedure using the following example.

Exercise 10.3: Use Examples 10.6 and 10.7 found in the *Statistics for Business and Economics* text.

Example 10.6: Find a 95% confidence interval for the mean monthly sales when the appliance store spends $400 on advertising.

Example 10.7: Predict the monthly sales for the next month, if $400 is spent on advertising. Use a 95% prediction interval.

Solution:

We solve Exercise 10.3 utilizing the **Simple Regression** function presented in the DDXL **Regression** menu (see Figure 10.10). **Open** the Data File **Adsales** by following the directions found in the preface of this manual. If done correctly, the data should appear in a workbook similar to that shown below in Figure 10.11. Use the mouse to select the data shown in the workbook.

Figure 10.10

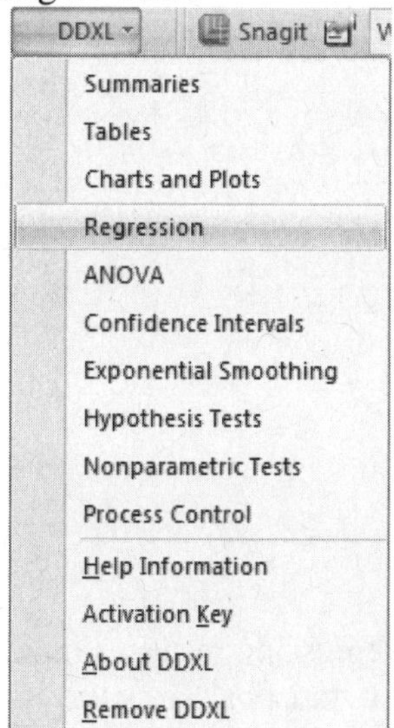

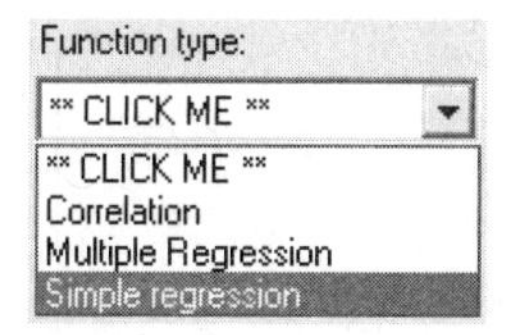

Click on the **DDXL Add-In** menu. Click on the **Regression** option and click on the in the Function Type box to access the different procedures available to select. Highlight the **Simple Regression** option to access the Regression Dialog menu shown in Figure 10.12. Highlight the **SALES** variable found in the **Names and Columns Box** and click on the to the right of the **Response Variable** box. Next, we highlight the **ADVERTISING** variable found in the **Names and Columns Box** and click on the to the right of the **Explanatory Variable** box. Click **OK**. DDXL gives us the output shown below in Figure 10.13.

Figure 10.11

	A	B
1	ADVERTISING	SALES
2	1	1
3	2	1
4	3	2
5	4	2
6	5	4

Figure 10.12

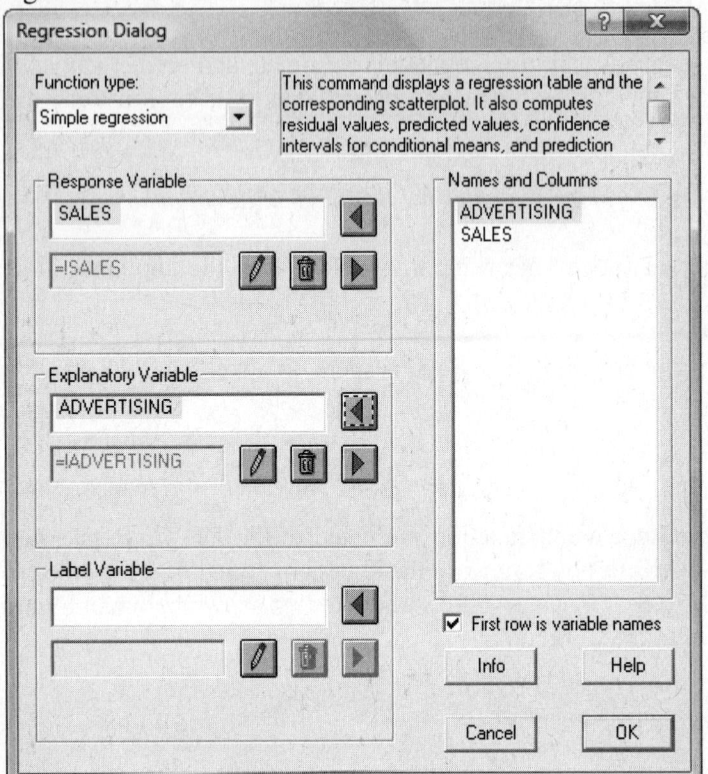

Figure 10.13

Dependent variable is: **SALES**
No Selector
R squared = 81.7% R squared (adjusted) = 75.6%
s = 0.6055 with 5 - 2 = 3 degrees of freedom

Source	Sum of Squares	df	Mean Square	F-ratio
Regression	4.9	1	4.9	13.4
Residual	1.1	3	0.366667	

Variable	Coefficient	s.e. of Coeff	t-ratio	prob
Constant	-0.1	0.6351	-0.157	0.8849
ADVERTIS...	0.7	0.1915	3.66	0.0354

DDXL provides several different types of output as part of its standard Simple Regression results. Figure 10.13 shows the basic simple linear regression output that is provided by most statistical software. In order to create the 95% confidence and prediction intervals, we must click on the **95% Confidence and Prediction Intervals** box shown in Figure 10.14 below. The resulting output is shown in Figure 10.15.

Figure 10.14

A regression is a formal summary of the linear relationship between two quantitative variables. You should always look at a scatterplot of the two variables to ensure that the underlying relationship is linear.

The regression calculates a straight line equation of the form y = a + bx. The coefficients a and b are the constant coefficient and slope coefficient respectively. The constant coefficient is reported in the first row of the Coefficient column at the bottom of the regression table. The slope coefficient is reported in the second row of the coefficient column at the bottom of the regression table.

The R-squared statistic, also called the coefficient of determination, measures the fraction of variability of y accounted for by its least squares linear regression on x. R-squared is always between 0 and 1. A value close to one indicates that there is a strong linear relationship to the data. A value close to 0 indicates that there is a weak relationship or no relationship. R-squared is reported at the top of the regression table.

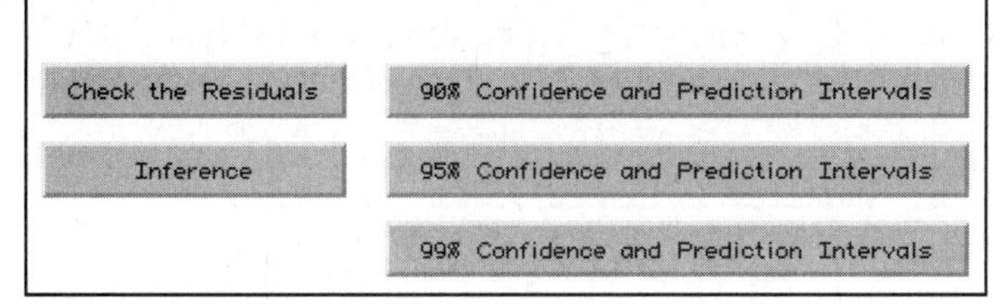

Figure 10.15

SALES	ADVERTISING	Predicted Value	Lower Cond. Mean Limit	Upper Cond. Mean Limit	Lower Prediction Limit	Upper Prediction Limit
1	1	0.6	-0.89270095	2.0927009	-1.8375704	3.0375704
1	2	1.3	0.24450104	2.355499	-0.8971963	3.4971963
2	3	2	1.1381887	2.8618113	-0.11099793	4.1109979
2	4	2.7	1.644501	3.755499	0.5028037	4.8971963
4	5	3.4	1.9072991	4.8927009	0.96242956	5.8375704

We find both the confidence interval for *E(y)* and the prediction interval for *y* by locating the row when advertising expenditure is $400 (i.e., Advertising = 4). We can see from the printout that the 95% confidence interval for E(y) is given as (1.645, 3.755) and the 95% prediction interval for y is given as (0.503, 4.897). Compare these intervals to the ones found in the text.

Technology Lab

The following exercise from the *Statistics for Business and Economics* text is given for you to practice the procedure covered in the text that is available within DDXL. Included with the exercise is the DDXL output that was generated to solve the problem.

10.88 Emotional exhaustion, or *burnout*, is a significant problem for people with careers in the field of human services. Regression analysis was used to investigate the relationship between burnout and aspects of the human service's professional's job and job-related behavior (*Journal of Applied Behavioral Science*, Vol. 22, 1986). Emotional exhaustion was measured with the Maslach Burnout Inventory, a questionnaire. One of the independent variables considered, called *concentration*, was the proportion of social contacts with individuals who belong to a person's work group. The table below lists the values of the emotional exhaustion index (higher values indicate greater exhaustion) and concentration for a sample of 25 human services professionals who work in a large public hospital.

Exhaustion Index, y	Concentration, x	Exhaustion Index, y	Concentration, x
100	20	493	86
525	60	892	83
300	38	527	79
980	88	600	75
310	79	855	81
900	87	709	75
410	68	791	77
296	12	718	77
120	35	684	77
501	70	141	17
920	80	400	85
810	92	970	96
506	77		

a. Construct a scattergram for the data. Do the variables *x* and *y* appear to be related?

b. Find the correlation coefficient for the data and interpret its value. Does your conclusion mean that concentration causes emotional exhaustion? Explain.

c. Test the usefulness of the straight-line relationship with concentration for predicting burnout. Use $\alpha = .05$.

d. Find the coefficient of determination for the model and interpret it.

e. Find a 95% confidence interval for the slope β_1. Interpret the result.

f. Use a 95% confidence interval to estimate the mean exhaustion level for all professionals who have 80% of their social contacts within their work groups. Interpret the interval.

DDXL Output

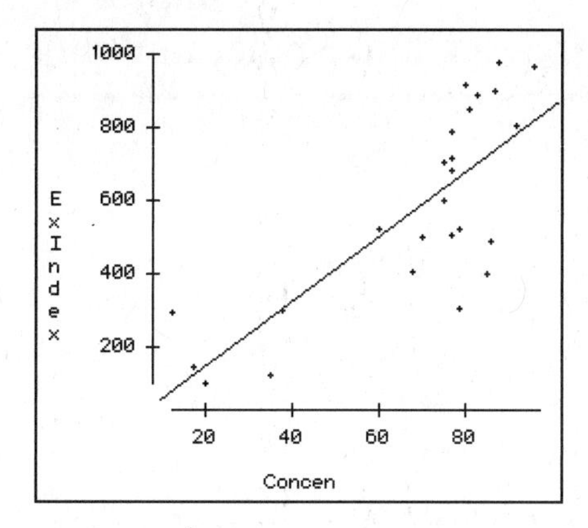

Correlation	0.783

Dependent variable is: **ExIndex**
No Selector
R squared = 61.2%   R squared (adjusted) = 59.5%
s = 174.2 with 25 - 2 = 23 degrees of freedom

Source	**Sum of Squares**	**df**	**Mean Square**	**F-ratio**
Regression	1.10241e6	1	1.10241e6	36.3
Residual	698009	23	30348.2	

Variable	**Coefficient**	**s.e. of Coeff**	**t-ratio**	**prob**
Constant	-29.4967	106.7	-0.276	0.7847
Concen	8.86547	1.471	6.03	≤ 0.0001

ExIndex	**Concen**	**Predicted Value**	**Lower Cond. Mean Limit**	**Upper Cond. Mean Limit**	**Lower Prediction Limit**	**Upper Prediction Limit**
100	20	147.81271	-16.591089	312.21651	-248.29222	543.91764
525	60	502.43157	425.79429	579.06884	133.99734	870.86579
300	38	307.39119	189.73883	425.04356	-71.703236	686.48563
980	88	750.66476	657.42323	843.9063	378.42225	1122.9073
310	79	670.87552	592.10998	749.64106	301.99272	1039.7583
900	87	741.79929	650.45792	833.14067	370.02819	1113.5704
410	68	573.35534	501.26009	645.45058	205.83903	940.87164
296	12	76.888939	-109.69937	263.47724	-328.92592	482.7038
120	35	280.79478	155.80209	405.78747	-100.64156	662.23112
501	70	591.08628	518.87811	663.29445	223.54781	958.62475
920	80	679.74099	599.69975	759.78223	310.5837	1048.8983
810	92	786.12665	684.72588	887.52742	411.75701	1160.4963
506	77	653.14458	576.63063	729.65852	284.73598	1021.5532
493	86	732.93382	643.42947	822.43817	361.60979	1104.2579
892	83	706.33741	621.92484	790.74998	336.20771	1076.4671
527	79	670.87552	592.10998	749.64106	301.99272	1039.7583
600	75	635.41364	560.72206	710.10521	267.37921	1003.4481
855	81	688.60646	607.19572	770.01721	319.14981	1058.0631
709	75	635.41364	560.72206	710.10521	267.37921	1003.4481
791	77	653.14458	576.63063	729.65852	284.73598	1021.5532
718	77	653.14458	576.63063	729.65852	284.73598	1021.5532
684	77	653.14458	576.63063	729.65852	284.73598	1021.5532
141	17	121.2163	-51.438538	293.87113	-278.38375	520.81635
400	85	724.06835	636.33393	811.80277	353.16697	1094.9697
970	96	821.58853	711.28649	931.89058	444.71051	1198.4666

Chapter 11
Multiple Regression and Model Building

11.1 Introduction

Chapter 11 in *Statistics for Business and Economics* introduces the topic of **multiple** regression analysis to the reader. While Chapter 10 served as the introduction to the general concepts of simple linear regression, Chapter 11 expands these concepts to modeling with several variables. In addition, Chapter 11 examines some of the problems associated with regression analysis and gives methods of detecting and solving these problems.

We utilize Chapter 11 examples to build on the linear regression base developed in the preceding chapter. Through the use of the Regression data analysis tool, DDXL allows the user to build more sophisticated models than the linear models of Chapter 10. We examine both the model building methods and the residual analysis options offered within DDXL. We will use the chapter examples that are given in the text to illustrate these methods. The following examples from *Statistics for Business and Economics* are solved with DDXL in this chapter:

Excel Companion Exercise	Page	Statistics for Business and Economics Example	Excel File Name
11.1	158	Example 11.3	GFClocks
11.2	161	Example 11.4	GFClocks
11.3	164	Example 11.6	GFClocks
11.4	166	Example 11.12	GFClocks
11.5	168	Throughout	Residual Analysis

11.2 Multiple Regression Model Building

We have seen in Chapter 10 how to use DDXL to build a simple linear regression model using one independent variable, x. The next step in our regression process in to add more independent variables into the regression model. DDXL allows for this using the same menus as seen in the simple linear regression chapter. We use an example from the text below.

Exercise 11.1: We use Example 11.3 found in the *Statistics for Business and Economics* text.

A collector of antique grandfather clocks knows that the price (y) received for the clocks increases linearly with the age (x_1) of the clocks. Moreover, the collector hypothesizes that the auction price (y) of the clocks will increase linearly as the number of bidders (x_2) increases. Thus, the following model is hypothesized:

$$y = \beta_0 + \beta_1 x_1 + \beta_2 x_2 + \varepsilon$$

A sample of 32 auction prices of grandfather clocks, along with their age and the number of bidders is shown in Table 11.1 Use DDXL to fit the model, $y = \beta_0 + \beta_1 x_1 + \beta_2 x_2 + \varepsilon$, and answer the following question:

a. Test the hypothesis that the mean auction price of a clock increases as the number of bidders increases when age is held constant, that is, $\beta_2 > 0$. Use $\alpha = .05$.

Table 11.1

Age(x_1)	Number of Bidders (x_2)	Auction Price (y)	Age(x_1)	Number of Bidders (x_2)	Auction Price (y)
127	13	1,235	170	14	2,131
115	12	1,080	182	8	1,550
127	7	845	162	11	1,884
150	9	1,522	184	10	2,041
156	6	1,047	143	6	845
182	11	1,979	159	9	1,483
156	12	1,822	108	14	1,055
132	10	1,253	175	8	1,545
137	9	1,297	108	6	729
113	9	946	179	9	1,792
137	15	1,713	111	15	1,175
117	11	1,024	187	8	1,593
137	8	1,147	111	7	785
153	6	1,092	115	7	744
117	13	1,152	194	5	1,356
126	10	1,336	168	7	1,262

Solution:

We solve Exercise 11.1 utilizing the **Multiple Regression** function presented in the DDXL **Regression** menu (see Figure 11.1). **Open** the Data File **GFClocks** by following the directions found in the preface of this manual. If done correctly, the data should appear in a workbook similar to that shown below in Figure 11.2. Use the mouse to select the data shown in the workbook.

Figure 11.1

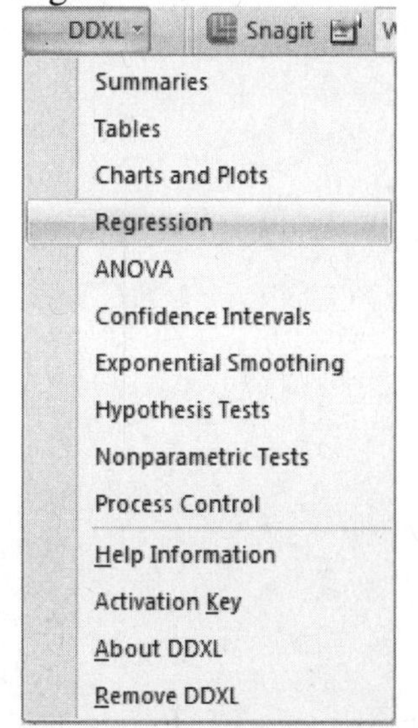

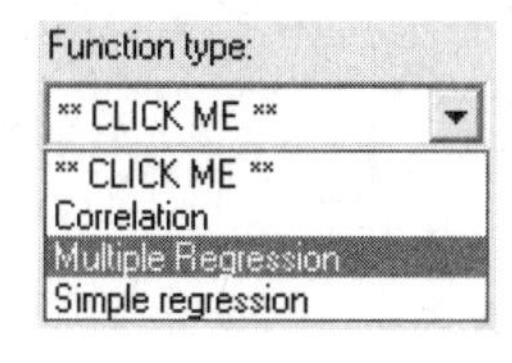

Figure 11.2

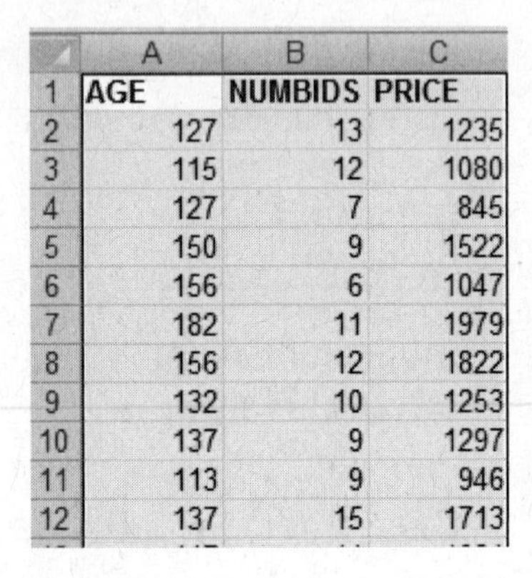

	A	B	C
1	AGE	NUMBIDS	PRICE
2	127	13	1235
3	115	12	1080
4	127	7	845
5	150	9	1522
6	156	6	1047
7	182	11	1979
8	156	12	1822
9	132	10	1253
10	137	9	1297
11	113	9	946
12	137	15	1713

Click on the **DDXL Add-In** menu. Click on the **Regression** option and click on the in the Function Type box to access the different procedures available to select. Highlight the **Multiple Regression** option to access the Regression Dialog menu shown in Figure 11.3. Highlight the **PRICE** variable found in the **Names and Columns Box** and click on the to the right of the **Response Variable** box. Next, we highlight the **AGE** variable found in the **Names and Columns Box** and click on the to the right of the **Explanatory Variable** box. We also highlight the **NUMBIDS** variable found in the **Names and Columns Box** and click on the to the right of the **Explanatory Variable** box. Click **OK**. DDXL gives us the output shown below in Figure 11.4.

Figure 11.3

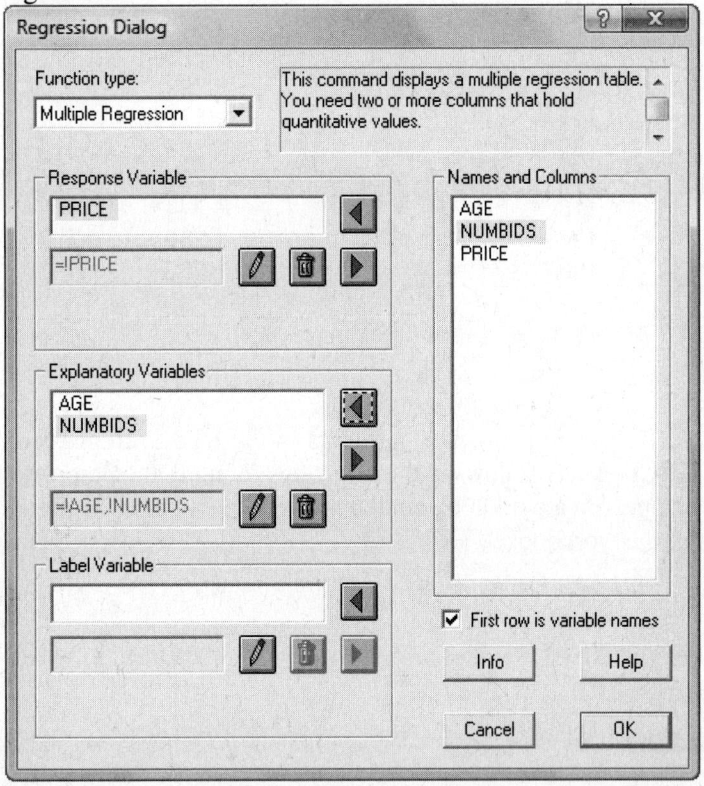

Figure 11.4

Dependent variable is: PRICE
No Selector
R squared = 89.2% R squared (adjusted) = 88.5%
s = 133.5 with 32 - 3 = 29 degrees of freedom

Source	Sum of Squares	df	Mean Square	F-ratio
Regression	4.28306e6	2	2.14153e6	120
Residual	516727	29	17818.2	

Variable	Coefficient	s.e. of Coeff	t-ratio	prob
Constant	-1338.95	173.8	-7.7	≤ 0.0001
AGE	12.7406	0.9047	14.1	≤ 0.0001
NUMBIDS	85.953	8.729	9.85	≤ 0.0001

Figure 11.4 shows the typical regression output provided by DDXL. In addition to the test for β_2 desired in this exercise, DDXL provides many other useful regression results. We point out the most important features of the printout generated by DDXL below. Compare these values to those shown in the *Statistics for Business and Economics* text.

DDXL Printout Values	**Description of Values**
R Square = 89.2%	Coefficient of determination
s = 133.5	Standard Deviation
Constant Coefficient = -1338.95	Estimate of β_0
AGE Coefficient = 12.7406	Estimate of β_1
NUMBIDS Coefficient = 85.953	Estimate of β_2
t-ratio for NUMBIDS = 9.85	Test Statistic for testing β_2
P-value for NUMBIDS = 0.0001	P-value for testing β_2
t-ratio for AGE = 14.1	Test Statistic for testing β_1
P-value for AGE = 0.0001	P-value for testing β_1
F-ratio = 120	Global F Test Statistic

We find the t-value for the test of β_2 is equal to t = 9.85 and the p-value for the desired test is p = .0001. The estimates of the β coefficients can be found in the Coefficients column in the printout. Our estimates of β_0, β_1, and β_2 are -1338.95, 12.74 and 85.95, respectively. We refer you to the text for more detailed information regarding the interpretations and conclusion that should be made for these values.

The next step of a regression analysis is to test all the hypothesized variables simultaneously. We refer to this process as checking the usefulness of the model. This process is illustrated in the following example.

Exercise 11.2: We use Example 11.4 found in the *Statistics for Business and Economics* text.

A collector of antique grandfather clocks knows that the price received for the clocks increases linearly with the age of the clocks. Moreover, the collector hypothesizes that the auction price of the clocks will increase linearly as the number of bidders increase. Thus, the following model is hypothesized:

$$y = \beta_0 + \beta_1 x_1 + \beta_2 x_2 + \varepsilon$$

A sample of 32 auction prices of grandfather clocks, along with their age and the number of bidders is shown in Table 11.1 The model $y = \beta_0 + \beta_1 x_1 + \beta_2 x_2 + \varepsilon$ is fit to the data. Use DDXL to :

a. Find and interpret the adjusted coefficient of determination, R_a^2.

b. Conduct the global F-test of model usefulness at the $\alpha = .05$ level of significance.

Solution:

We need to generate the multiple regression model hypothesized above using DDXL. The printout generated must include the adjusted coefficient of determination and the global-F test and p-value information. Fortunately, the standard DDXL regression output yields both of the desired values.

We solve Exercise 11.2 utilizing the **Multiple Regression** function presented in the DDXL **Regression** menu (see Figure 11.5). **Open** the Data File **GFClocks** by following the directions found in the preface of this manual. If done correctly, the data should appear in a workbook similar to that shown below in Figure 11.6. Use the mouse to select the data shown in the workbook.

Figure 11.5

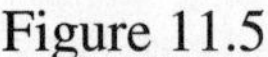

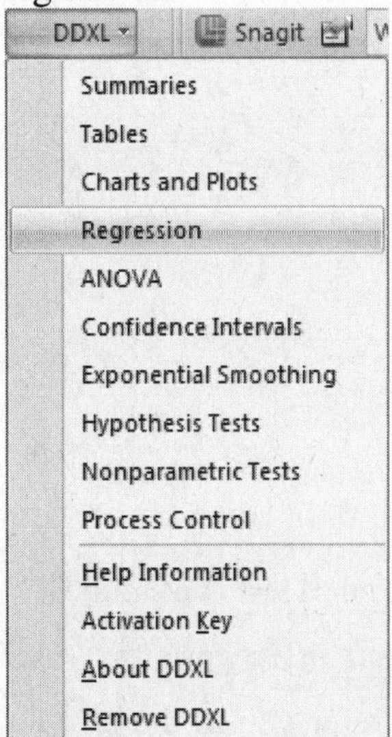

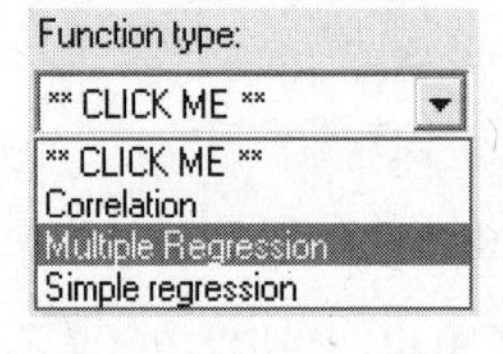

Figure 11.6

	A	B	C
1	AGE	NUMBIDS	PRICE
2	127	13	1235
3	115	12	1080
4	127	7	845
5	150	9	1522
6	156	6	1047
7	182	11	1979
8	156	12	1822
9	132	10	1253
10	137	9	1297
11	113	9	946
12	137	15	1713

Click on the **DDXL Add-In** menu. Click on the **Regression** option and click on the ▾ in the Function Type box to access the different procedures available to select. Highlight the **Multiple Regression** option to access the Regression Dialog menu shown in Figure 11.7. Highlight the **PRICE** variable found in the **Names and Columns Box** and click on the ◀ to the right of the **Response Variable** box. Next, we highlight the **AGE** variable found in the **Names and Columns Box** and click on the ◀ to the right of the **Explanatory Variable** box. We also highlight the **NUMBIDS** variable found in the **Names and Columns Box** and click on the ◀ to the right of the **Explanatory Variable** box. Click **OK**. DDXL gives us the output shown below in Figure 11.8.

Figure 11.7

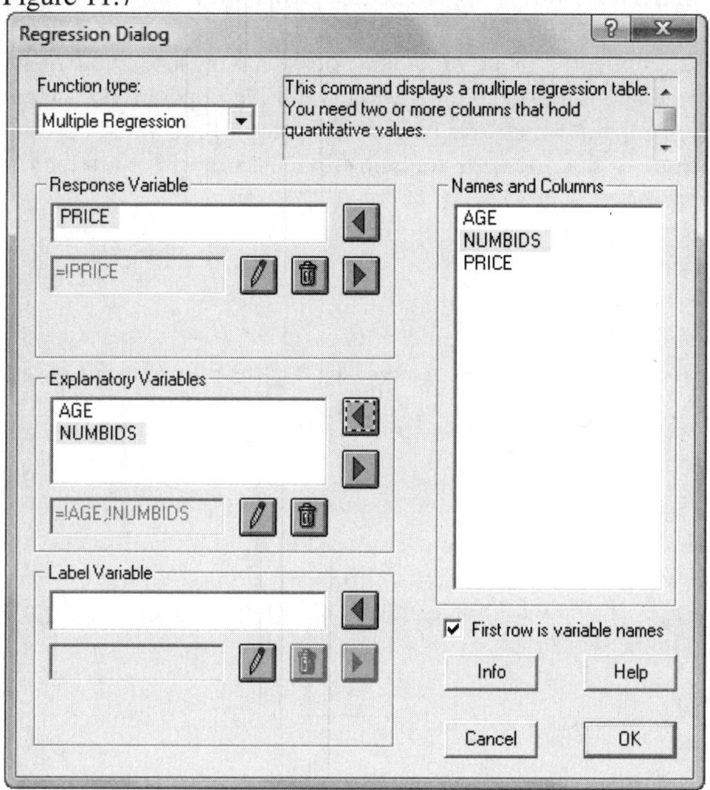

Figure 11.8

Dependent variable is: PRICE
No Selector
R squared = 89.2% R squared (adjusted) = 88.5%
s = 133.5 with 32 - 3 = 29 degrees of freedom

Source	Sum of Squares	df	Mean Square	F-ratio
Regression	4.28306e6	2	2.14153e6	120
Residual	516727	29	17818.2	

Variable	Coefficient	s.e. of Coeff	t-ratio	prob
Constant	-1338.95	173.8	-7.7	≤ 0.0001
AGE	12.7406	0.9047	14.1	≤ 0.0001
NUMBIDS	85.953	8.729	9.85	≤ 0.0001

The adjusted coefficient of determination R_a^2 is listed as the Adjusted R Square value in the Regression Statistics table above. The R_a^2 value of $R_a^2 = 88.5\%$ can be compared to the value shown in the text. The global F statistic is shown in the ANOVA table in the printout above. The global F statistic of F = 120 is identical to the value shown in the text. Unfortunately, DDXL does not provide a p-value for the F-test statistic and the user will need to compare this value with a critical value using the F-tables presented in the text or use a software package to generate the p-value corresponding to the test statistic. We refer you to the text for more detailed information regarding the interpretations and conclusions that should be made for these values.

The next step in the model building process of a regression analysis is to add various types of regression terms to the model. Whether the terms added are interactions, quadratics, or qualitative terms, the process within DDXL is the same. We illustrate this process by adding an interaction component to the preceding example to illustrate. Please note that the process of adding quadratic and qualitative terms to the regression model is identical to the process demonstrated in the next example.

Exercise 11.3: We use Example 11.6 found in the *Statistics for Business and Economics* text.

Refer to Examples 11.3 and 11.4. Suppose the collector of grandfather clocks, having observed many auctions, believes that the *rate of increase* of the auction price with age will be driven upward by a large number of bidders. Thus, instead of a relationship in which the rate of the price is the same for any number of bidders, the collector believes the slope of the price-age relationship increases as the number of bidders increases. Consequently, the interaction model is proposed:

$$Y = \beta_0 + \beta_1 x_1 + \beta_2 x_2 + \beta_3 x_1 x_2 + \varepsilon$$

a. Test the overall utility of the model using the global F-test at $\alpha = .05$.

b. Test the hypothesis (at $\alpha = .05$) that the price-age slope increases as the number of bidders increases - that is, that age and number of bidders, x_2, interact positively.

Solution:

We solve Exercise 11.3 utilizing the **Multiple Regression** function presented in the DDXL **Regression** menu (see Figure 11.9). **Open** the Data File **GFClocks** by following the directions found in the preface of this manual. If done correctly, the data should appear in a workbook similar to that shown below in Figure 11.10. Use the mouse to select the data shown in the workbook.

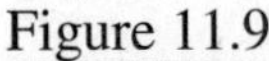
Figure 11.9

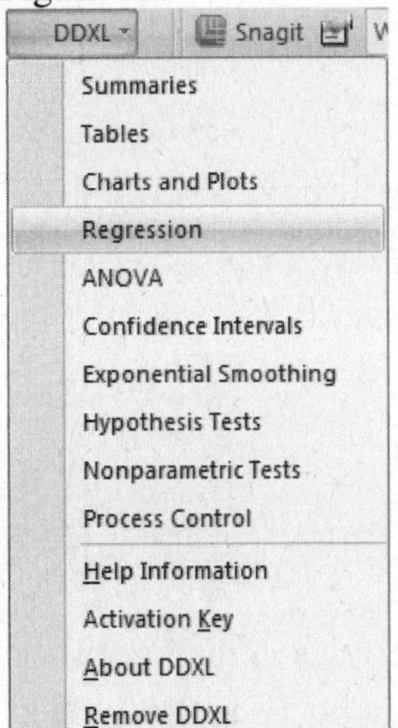

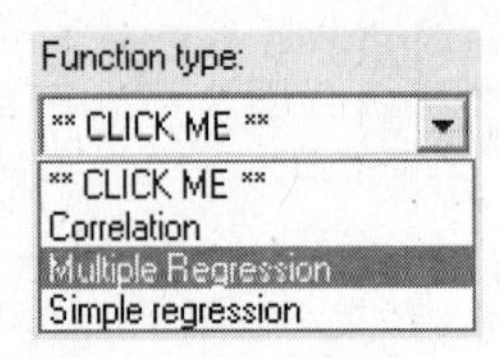

Figure 11.10

	A	B	C	D
1	AGE	NUMBIDS	PRICE	AGE-BID
2	127	13	1235	1651
3	115	12	1080	1380
4	127	7	845	889
5	150	9	1522	1350
6	156	6	1047	936
7	182	11	1979	2002
8	156	12	1822	1872
9	132	10	1253	1320
10	137	9	1297	1233
11	113	9	946	1017
12	137	15	1713	2055
13	117	11	1024	1287

Click on the **DDXL Add-In** menu. Click on the **Regression** option and click on the ▾ in the Function Type box to access the different procedures available to select. Highlight the **Multiple Regression** option to access the Regression Dialog menu shown in Figure 11.11. Highlight the **PRICE** variable found in the **Names and Columns Box** and click on the ◀ to the right of the **Response Variable** box. Next, we highlight the **AGE** variable found in the **Names and Columns Box** and click on the ◀ to the right of the **Explanatory Variable** box. We also highlight the **NUMBIDS** variable found in the **Names and Columns Box** and click on the ◀ to the right of the **Explanatory Variable** box. To include the interaction term in the model, we also highlight the **AGE-BID** variable found in the **Names and Columns Box** and click on the ◀ to the right of the **Explanatory Variable** box. Click **OK**. DDXL gives us the output shown below in Figure 11.12.

Figure 11.11

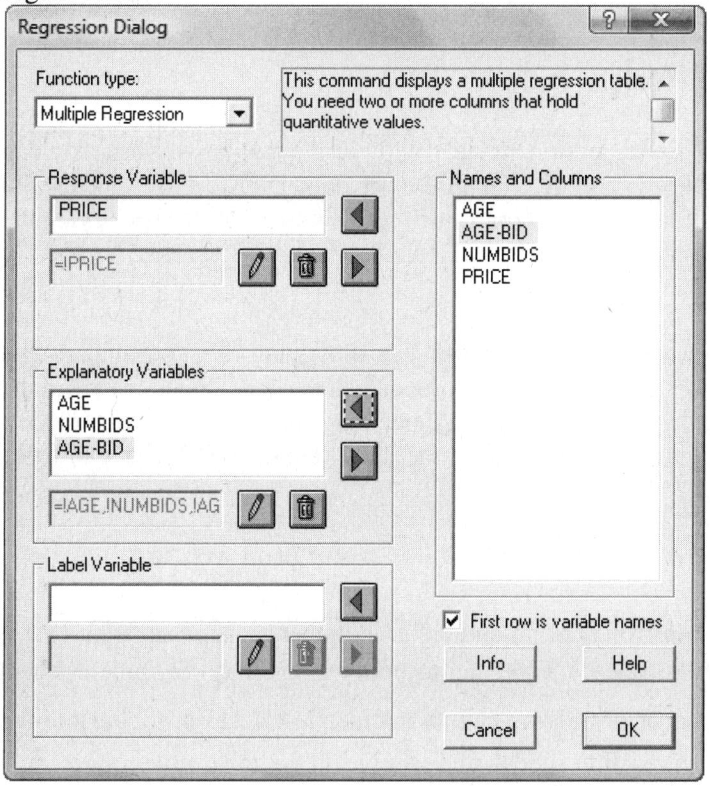

Figure 11.12

Dependent variable is: PRICE
No Selector
R squared = 95.4% R squared (adjusted) = 94.9%
s = 88.91 with 32 - 4 = 28 degrees of freedom

Source	Sum of Squares	df	Mean Square	F-ratio
Regression	4.57843e6	3	1.52614e6	193
Residual	221362	28	7905.79	

Variable	Coefficient	s.e. of Coeff	t-ratio	prob
Constant	320.458	295.1	1.09	0.2868
AGE	0.878142	2.032	0.432	0.6690
NUMBIDS	-93.2648	29.89	-3.12	0.0042
AGE-BID	1.29785	0.2123	6.11	≤ 0.0001

The global F-test is shown on the printout as F-ratio = 193 and the t-test for the interaction is shown as t-ratio = 6.11. Compare these values to the ones shown in the text.

11.3 Comparing Two Regression Models

We have seen how DDXL can be used to fit regression models with just quantitative variables and regression models with just qualitative variables. For more complicated models, DDXL allows the user to input both quantitative and qualitative variables into a single multiple regression model. By specifying the appropriate columns in the Regression data analysis menu as Explanatory Variables, any number of quantitative and qualitative variables can be combined.

The final step in the model building topic is to develop a method that allows the user to compare two regression models to determine which is the better predictor of the dependent variable. Section 11.9 in the text details the partial-F test for testing a portion of the regression model. By fitting two separate models within DDXL, it is possible to calculate the partial-F test statistic that the book details. We demonstrate with the following example.

Exercise 11.4: We use Example 11.12 found in the *Statistics for Business and Economics* text.

Refer to the problem of modeling the auction price (y) of an antique grandfather clock, Examples 11.1 – 11.6. In these examples we discovered that the price was related to both the age (x_1) of the clock and number of bidders (x_2), and the age and number of bidders had an interactive effect on price. Both the first-order model of Example 11.1 and the interaction model of Example 11.6, however, propose only straight-line (linear) relationships. We did not consider the possibility that the relationship between price (y) and age (x_1) is curvilinear, or that the relationship between price (y) and the number of bidders (x_2) is curvilinear.

a. Propose a complete 2nd-order model for price (y) as a function of age (x_1) and number of bidders (x_2).

b. Fit the model to the data for the 32 clocks and give the least squares prediction equation.

c. Do the data provide sufficient evidence to indicate that the quadratic terms in the model contribute information for the prediction of price (y)? That is, is there evidence of curvature in the price-age and price-bidders relationships?

Solution:

In order to compare the determine if the second-order terms contribute information for predicting *y*, both a full model (containing the second-order terms) and a reduced model (that does not contain the second-order terms) must be fit in DDXL. We utilize the procedures covered in the last section to fit both models (using Data File **GFClocks**). The corresponding regression output for both models is shown below in Figures 11.13 and 11.14.

Figure 11.13

Dependent variable is: **PRICE**
No Selector
R squared = 96.0% R squared (adjusted) = 95.2%
s = 86.1 with 32 - 6 = 26 degrees of freedom

Source	Sum of Squares	df	Mean Square	F-ratio
Regression	4.60704e6	5	921408	124
Residual	192752	26	7413.53	

Variable	Coefficient	s.e. of Coeff	t-ratio	prob
Constant	-331.926	764.9	-0.434	0.6679
AGE	3.20792	8.947	0.359	0.7228
Age Squared	-0.00299519	0.02748	-0.109	0.9140
NUMBIDS	14.8089	62.21	0.238	0.8137
Bids Squared	-4.17903	2.145	-1.95	0.0623
AGE-BID	1.12316	0.2316	4.85	≤ 0.0001

Figure 11.14

Dependent variable is: **PRICE**
No Selector
R squared = 95.4% R squared (adjusted) = 94.9%
s = 88.91 with 32 - 4 = 28 degrees of freedom

Source	Sum of Squares	df	Mean Square	F-ratio
Regression	4.57843e6	3	1.52614e6	193
Residual	221362	28	7905.79	

Variable	Coefficient	s.e. of Coeff	t-ratio	prob
Constant	320.458	295.1	1.09	0.2868
AGE	0.878142	2.032	0.432	0.6690
NUMBIDS	-93.2648	29.89	-3.12	0.0042
AGE-BID	1.29785	0.2123	6.11	≤ 0.0001

Compare these two printouts versus the MINITAB printouts found in the text. We refer you to the text for more detailed information regarding the interpretations and conclusion that should be made for these values.

11.4 Residual Analysis

So far we have covered model building and model testing within the DDXL program. The last topic to address is the topic of residual analysis. As specified in the text, the topic of residual analysis requires the construction of several different graphical displays that are readily available from the multiple regression menu within DDXL. We illustrate how to generate these plots using the following example from the text.

Exercise 11.5: We use the data from the grandfather clocks data that is used throughout this chapter to generate the plots necessary to complete a residual analysis.

Solution:

The data for the grandfather clock example used throughout this chapter are repeated in Table 11.2 in the text (and below in Table 11.7), with one important difference: The auction price of the clock at the top of the second column has been changed from $2,131 to 1,131 (saved as Data file **Residual Analysis**). The interaction model

$$y = \beta_0 + \beta_1 x_1 + \beta_2 x_2 + \beta_3 x_1 x_2 + \varepsilon$$

is again fit to these (modified) data. Use DDXL to generate all corresponding residual analysis printouts.

Table 11.2

Age(x_1)	Number of Bidders (x_2)	Auction Price (y)	Age(x_1)	Number of Bidders (x_2)	Auction Price (y)
127	13	1,235	170	14	2,131
115	12	1,080	182	8	1,550
127	7	845	162	11	1,884
150	9	1,522	184	10	2,041
156	6	1,047	143	6	845
182	11	1,979	159	9	1,483
156	12	1,822	108	14	1,055
132	10	1,253	175	8	1,545
137	9	1,297	108	6	729
113	9	946	179	9	1,792
137	15	1,713	111	15	1,175
117	11	1,024	187	8	1,593
137	8	1,147	111	7	785
153	6	1,092	115	7	744
117	13	1,152	194	5	1,356
126	10	1,336	168	7	1,262

We begin by building the interaction model as we did earlier in this chapter. The results from building the interaction model are shown again in Figure 11.15.

Figure 11.15

Dependent variable is: **PRICE**
No Selector
R squared = 95.4% R squared (adjusted) = 94.9%
s = 88.91 with 32 - 4 = 28 degrees of freedom

Source	Sum of Squares	df	Mean Square	F-ratio
Regression	4.57843e6	3	1.52614e6	193
Residual	221362	28	7905.79	

Variable	Coefficient	s.e. of Coeff	t-ratio	prob
Constant	320.458	295.1	1.09	0.2868
AGE	0.878142	2.032	0.432	0.6690
NUMBIDS	-93.2648	29.89	-3.12	0.0042
AGE-BID	1.29785	0.2123	6.11	≤ 0.0001

In addition to the regression output shown above, DDXL allows the user to click on the box, [Check Residuals] to generate residual analysis plots. The plots are shown below in Figures 11.16 and 11.17.

Figure 11.16

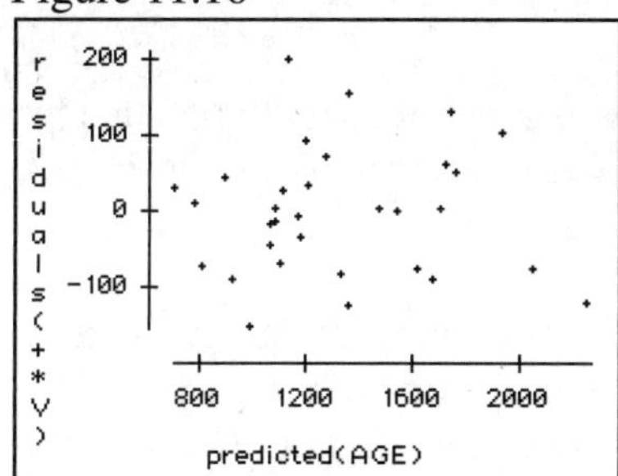

Figure 11.17

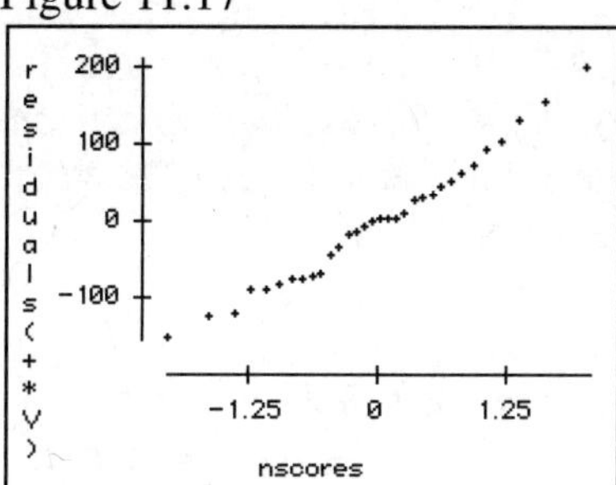

These plots can be used to check both the normal and equal variance assumptions in the residual analysis.

Technology Lab

The following exercise from the *Statistics for Business and Economics* text is given for you to practice the procedure covered in the text that is available within DDXL. Included with the exercise is the DDXL output that was generated to solve the problem.

11.154 A firm that has developed a new type of light bulb is interested in evaluating its performance in order to decide whether to market it. It is known that the light output of the bulb depends on the cleanliness of its surface area and the length of time the bulb has been in operation. Use the data in the table at right and the procedures you learned in this chapter to build a regression model that relates drop in light output to bulb surface cleanliness and length of operation. Be sure to conduct a residual analysis also.

DDXL Output

Dependent variable is: **Drop**
No Selector
R squared = 85.5% R squared (adjusted) = 82.9%
s = 5.391 with 14 - 3 = 11 degrees of freedom

Source	**Sum of Squares**	**df**	**Mean Square**	**F-ratio**
Regression	1886.66	2	943.33	32.5
Residual	319.696	11	29.0633	

Variable	**Coefficient**	**s.e. of Coeff**	**t-ratio**	**prob**
Constant	12.8036	2.97	4.31	0.0012
Length	0.0096875	0.001801	5.38	0.0002
Surface	-17.2857	2.882	-6	≤ 0.0001

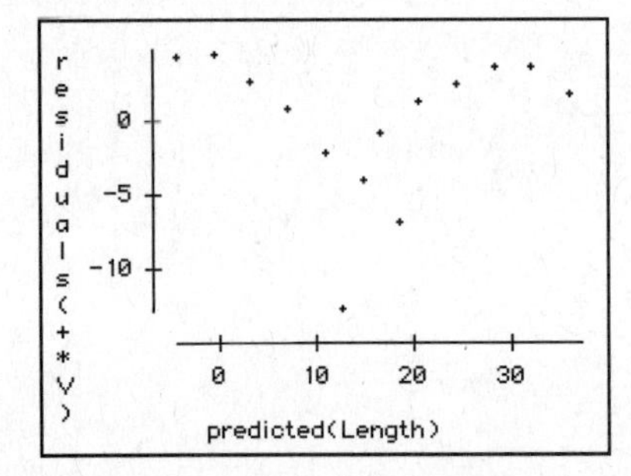

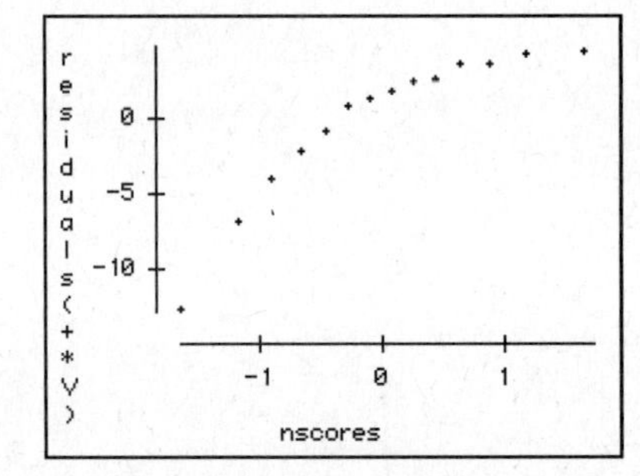

Chapter 12
Methods of Quality Improvement: Statistical Process Control

12.1 Introduction

Chapter 12 introduces the topic of quality improvement to the reader. The main topic covered in the text involves using scatter plots to create control charts that monitor the outcomes of a statistical process. These control charts specify upper and lower limits on the plot inside which the process is expected to stay. By using control charts, the user has a valid statistical tool that enables him/her to identify when a process is decreasing in quality. DDXL offers four different types of control charts for the user. They are the $\bar{x}$, R, s, and p-charts. The following examples from *Statistics for Business and Economics* are solved with DDXL in this chapter.

Excel Companion		Statistics for Business and Economics	
Exercise	**Page**	**Example**	**Excel File Name**
12.1	171	Example 12.6/12.8	Paint125
12.2	174	Example 12.9	Warehouse

12.2 Constructing R-Charts and $\bar{x}$-Charts

The DDXL program allows for construction of the $\bar{x}$-chart and R-chart when working with quantitative data. In order to create these charts the user must create columns of data for the individual values that are being measured and for the sample that the measurements were collected in. We illustrate these charts with the following example.

Exercise 12.1: We combine Examples 12.6 and Example 12.8 found in the *Statistics for Business and Economics* text.

Let's return to the paint-filling process described in Sections 12.2 and 12.3 of the text. Suppose instead of sampling 50 consecutive gallons of paint from the filling process to develop a control chart, it was decided to sample five consecutive cans once each hour for the next 25 hours. The sample data are presented in Table 12.1. This sampling strategy (rational subgrouping) was selected because several times a month the filling head in question becomes clogged. When that happens, the head dispenses less and less paint over the course of the day. However, the pattern of decrease is so irregular that minute-to-minute or even half-hour-to-half-hour changes are difficult to detect.

a. Construct an $\bar{x}$-chart for the process using the data below (from Example 12.6 in the text).
b. Construct an R-chart for the process using the data below (from Example 12.8 in the text).

Table 12.1

Sample	Measurements					Mean	Range
1	10.0042	9.9981	10.001	9.9964	10.0001	9.99995	0.0078
2	9.995	9.9986	9.9948	10.003	9.9938	9.99704	0.0092
3	10.0028	9.9998	10.0086	9.9949	9.998	10.00082	0.0137
4	9.9952	9.9923	10.0034	9.9965	10.0026	9.998	0.0111
5	9.9997	9.9983	9.9975	10.0078	9.9891	9.99649	0.0195
6	9.9987	10.0027	10.0001	10.0027	10.0029	10.00141	0.0042
7	10.0004	10.0023	10.0024	9.9992	10.0135	10.00358	0.0143
8	10.0013	9.9938	10.0017	10.0089	10.0001	10.00116	0.0151
9	10.0103	10.0009	9.9969	10.0103	9.9986	10.00339	0.0134
10	9.998	9.9954	9.9941	9.9958	9.9963	9.99594	0.0039
11	10.0013	10.0033	9.9943	9.9949	9.9999	9.99874	0.009
12	9.9986	9.999	10.0009	9.9947	10.0008	9.99882	0.0062
13	10.0089	10.0056	9.9976	9.9997	9.9922	10.0008	0.0167
14	9.9971	10.0015	9.9962	10.0038	10.0022	10.00016	0.0076
15	9.9949	10.0011	10.0043	9.9988	9.9919	9.99822	0.0124
16	9.9951	9.9957	10.0094	10.004	9.9974	10.00033	0.0137
17	10.0015	10.0026	10.0032	9.9971	10.0019	10.00127	0.0061
18	9.9983	10.0019	9.9978	9.9997	10.0029	10.0013	0.0051
19	9.9977	9.9963	9.9981	9.9968	10.0009	9.99798	0.0127
20	10.0078	10.0004	9.9966	10.0051	10.0007	10.00212	0.0112
21	9.9963	9.999	10.0037	9.9936	9.9962	9.99764	0.0101
22	9.9999	10.0022	10.0057	10.0026	10.0032	10.00272	0.0058
23	9.9998	10.0002	9.9978	9.9966	10.006	10.00009	0.0094
24	10.0031	10.0078	9.9988	10.0032	9.9944	10.00146	0.0134
25	9.9993	9.9978	9.9964	10.0032	10.0041	10.00015	0.0077

Solution:

We solve Exercise 12.1 utilizing the **Mu, R, s Control Charts** function presented in the DDXL **Process Control** menu (see Figure 12.2). **Open** the Data File **Paint125** which is found on the data disk included with this manual. If done correctly, the data should appear in a workbook similar to that shown below in Figure 12.1. Use the mouse to select the data shown in the workbook.

Figure 12.1

	A	B
1	Sample	Weight
2	1	10.0042
3	1	9.9981
4	1	10.001
5	1	9.9964
6	1	10.0001
7	2	9.995

Figure 12.2

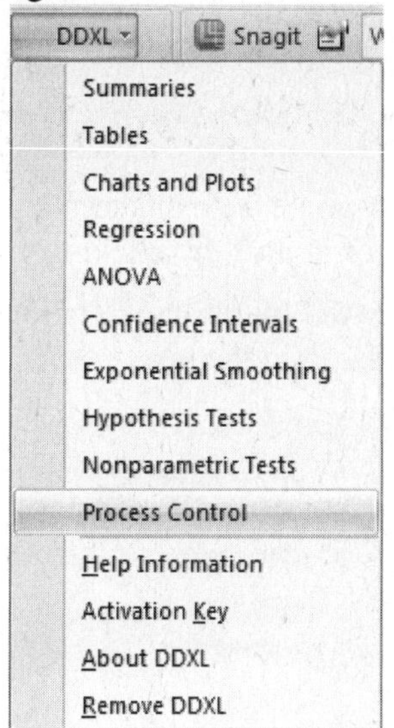

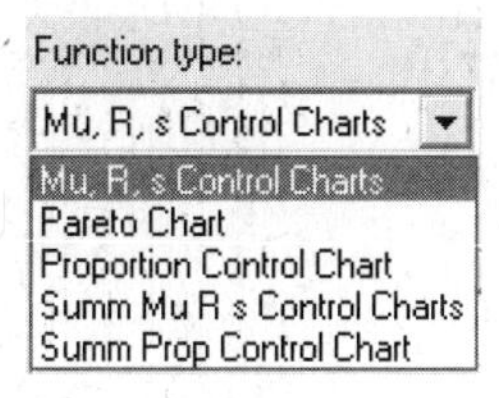

Click on the **DDXL Add-In** menu. Click on the **Process Control** option and click on the in the Function Type box to access the different procedures available to select. Highlight the **Mu, R, s Control Charts** option to access the Process Control Dialog menu shown in Figure 12.3. Highlight the **Weight** variable found in the **Names and Columns Box** and click on the to the right of the **Quantitative Variable** box. Next, we highlight the **Sample** variable found in the **Names and Columns Box** and click on the to the right of the **Group Variable** box. Click **OK**. DDXL provides us with the menu to create the charts that we want as shown in figure 12.4.

Figure 12.3

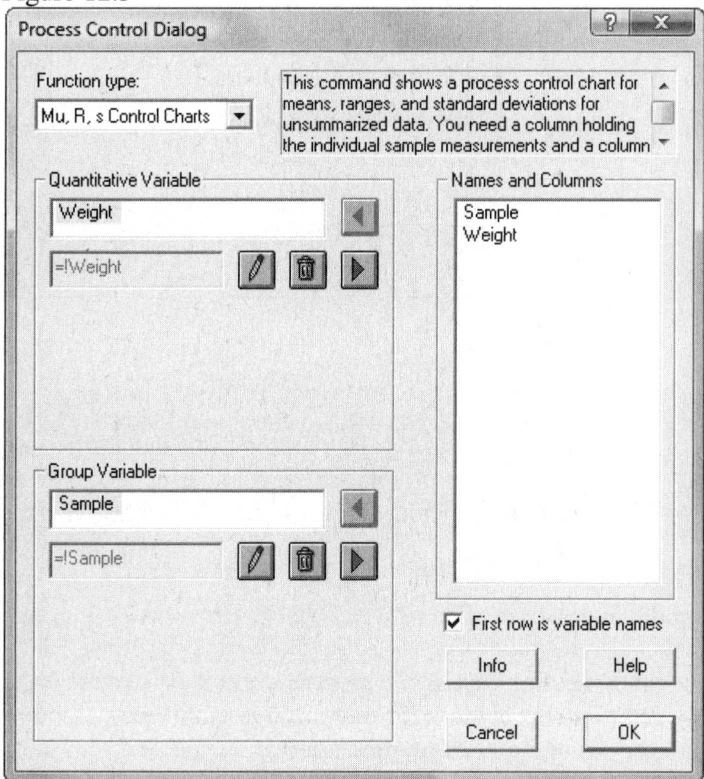

Figure 12.4

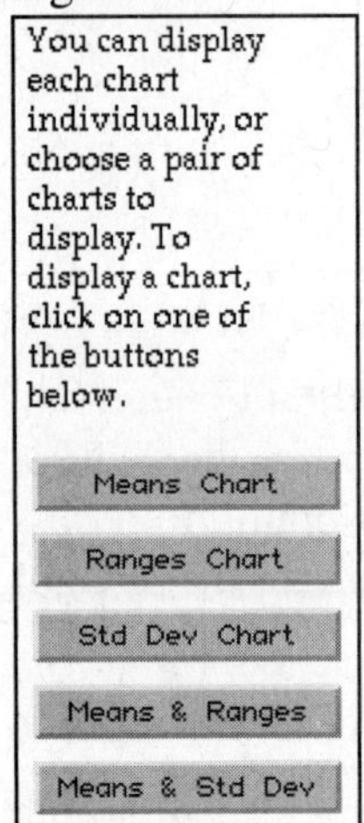

We create the chart that we desire by clicking on the appropriate box in the menu shown above. We can create the charts individually or together in a set by clicking on the 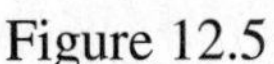

box. By clicking on this box, we get the two charts shown below in Figures 12.5 and 12.6. Compare these charts to the charts found in the text.

Figure 12.5

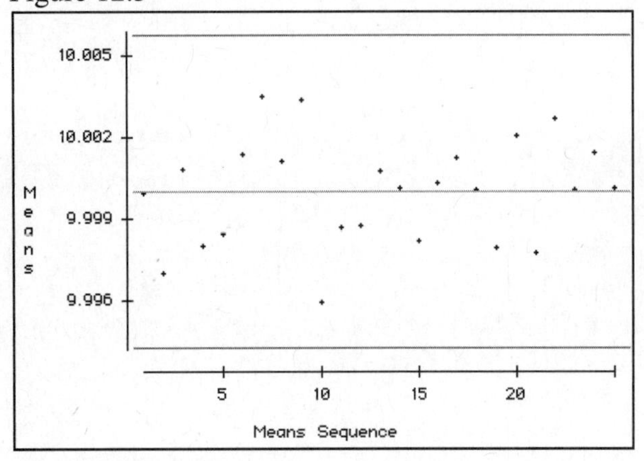

Figure 12.6

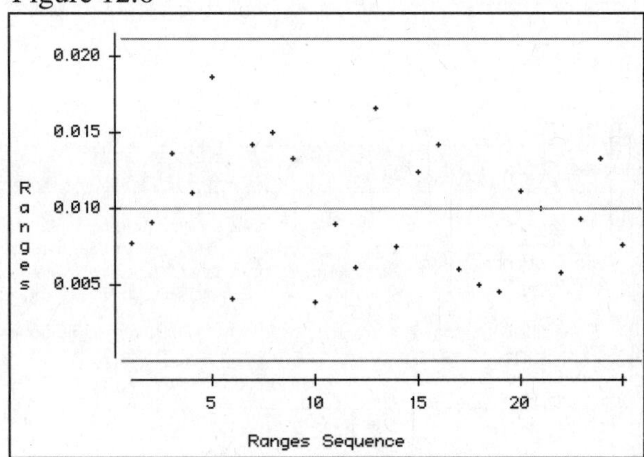

12.3 Constructing p-Charts

The DDXL program allows for construction of the p-chart for charting the proportion of defectives in a process. In order to create this chart, the user must create a column of data for the number of successes in a sample and a column of data for the total size of the sample collected. We illustrate the p-chart with the following example.

Exercise 12.2: We use Example 12.9 found in the *Statistics for Business and Economics* text.

A manufacturer of auto parts is interested in implementing statistical process control in several areas within its warehouse operation. The manufacturer wants to begin with the order assembly process. Too frequently orders received by customers contain the wrong items or too few items.

For each order received, parts are picked from storage bins in the warehouse, labeled, and placed on a conveyor belt system. Since the bins are spread over a three-acre area, items that are part of the same order may be placed on different spurs of the conveyor belt system. Near the end of the belt system all spurs converge and a worker sorts the items according to the order they belong to. That information is contained on labels that were placed on the items by the pickers.

The workers have identified three errors that cause shipments to be improperly assembled: (1) pickers pick from the wrong bin, (2) pickers mislabel items, and (3) the sorter makes an error.

The firm's quality manager has implemented a sampling program in which 90 assembled orders are sampled each day and checked for accuracy. An assembled order is considered nonconforming (defective) if it differs in any way from the order placed by the customer. To date, 25 samples have been evaluated. The resulting data are shown in Table 12.2. Construct a p-chart for the order assembly operation.

Table 12.2

Sample	Size	Defective Orders	Sample Proportion
1	90	12	0.13333
2	90	6	0.06667
3	90	11	0.12222
4	90	8	0.08889
5	90	13	0.14444
6	90	14	0.15556
7	90	12	0.13333
8	90	6	0.06667
9	90	10	0.11111
10	90	13	0.14444
11	90	12	0.13333
12	90	24	0.26667
13	90	23	0.25556
14	90	22	0.24444
15	90	8	0.08889
16	90	3	0.03333
17	90	11	0.12222
18	90	14	0.15556
19	90	5	0.05556
20	90	12	0.13333
21	90	18	0.20000
22	90	12	0.13333
23	90	13	0.14444
24	90	4	0.04444
25	90	6	0.06667
Totals	2,250	292	

Solution:

We solve Exercise 12.2 utilizing the **Summ Prop Control Chart** function presented in the DDXL **Process Control** menu (see Figure 12.8). **Open** the Data File **Warehouse** which is found on the data disk included with this manual. If done correctly, the data should appear in a workbook similar to that shown below in Figure 12.7. Use the mouse to select the data shown in the workbook.

Figure 12.7

	A	B	C	D
1	SAMPLE	N	DEFECTS	PDEFECT
2	1	90	12	0.133333
3	2	90	6	0.066667
4	3	90	11	0.122222
5	4	90	8	0.088889
6	5	90	13	0.144444
7	6	90	14	0.155556
8	7	90	12	0.133333

Figure 12.8

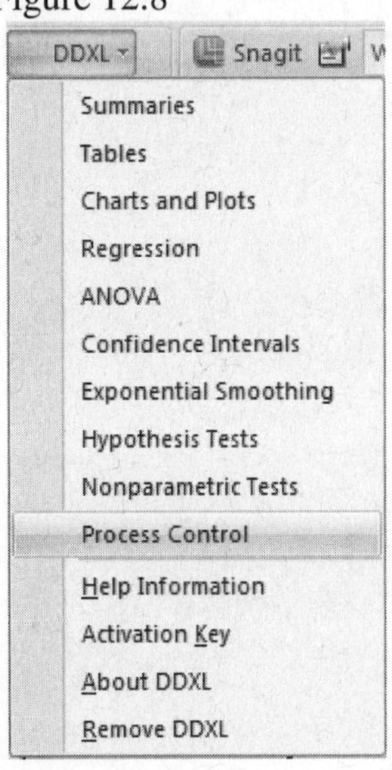

Click on the **DDXL Add-In** menu. Click on the **Process Control** option and click on the in the Function Type box to access the different procedures available to select. Highlight the **Summ Prop Control Chart** option to access the Process Control Dialog menu shown in Figure 12.9. Highlight the **DEFECTS** variable found in the **Names and Columns Box** and click on the to the right of the **Successes Variable** box. Next, we highlight the **N** variable found in the **Names and Columns Box** and click on the to the right of the **Totals Variable** box. Click **OK**. DDXL provides us with the menu to create the charts that we want as shown in figure 12.10.

Figure 12.9

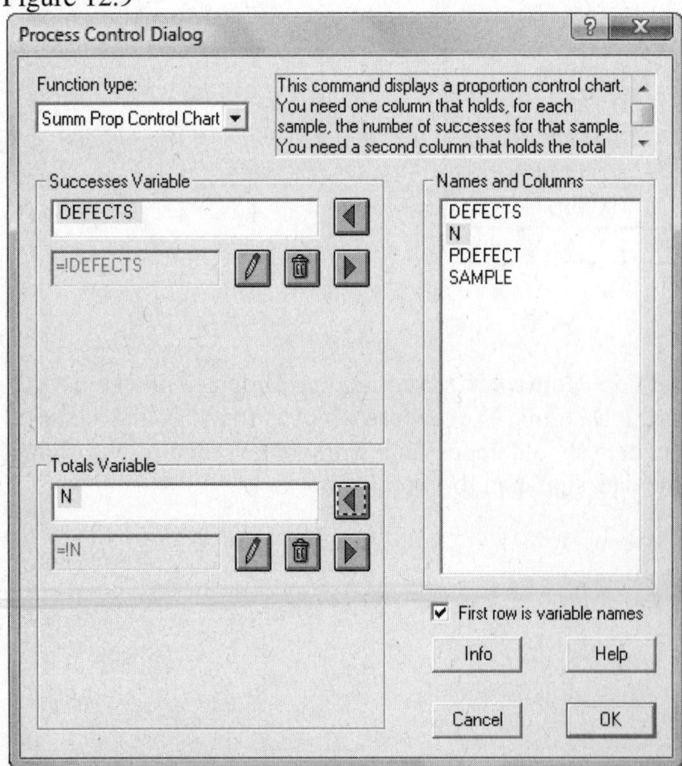

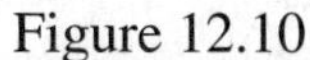
Figure 12.10

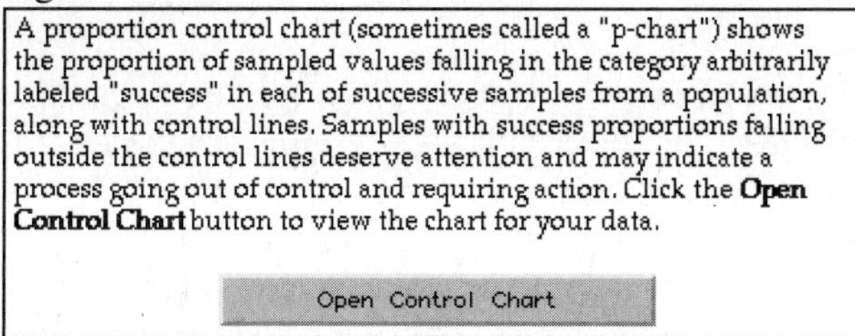
A proportion control chart (sometimes called a "p-chart") shows the proportion of sampled values falling in the category arbitrarily labeled "success" in each of successive samples from a population, along with control lines. Samples with success proportions falling outside the control lines deserve attention and may indicate a process going out of control and requiring action. Click the **Open Control Chart** button to view the chart for your data.

Open Control Chart

We create the p-chart that we desire by clicking on the Open Control Chart box. By clicking on this box, we get the chart shown below in Figures 12.11. Compare this chart to the one shown in the text.

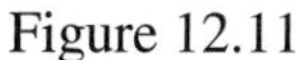
Figure 12.11

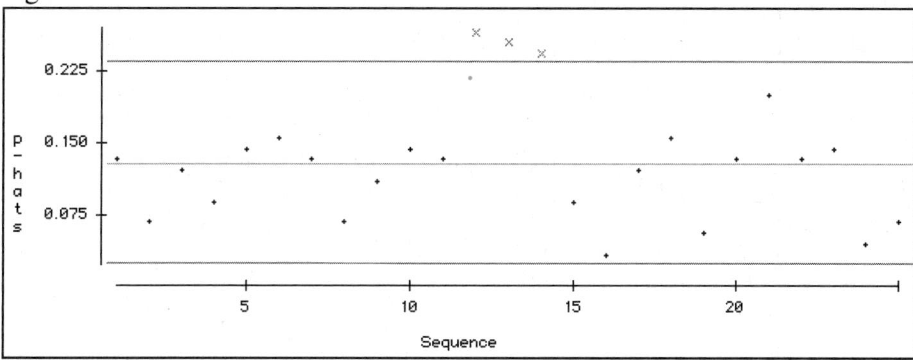

Technology Lab

The following exercises from the *Statistics for Business and Economics* text are given for you to practice the control chart procedures that are available within DDXL. Included with the exercises are the DDXL outputs that were generated to solve the problems.

12.77 Officials at Mountain Airlines are interested in monitoring the length of time customers wait in line to check in at their airport counter in Reno, Nevada. In order to develop a control chart, five customers were sampled each day for 20 days. The data, in minutes, are presented below (file SBE Exercise 12.73).

Sample	Waiting Times				
1	3.20	6.70	1.30	8.40	2.20
2	5.00	4.10	7.90	8.10	0.40
3	7.10	3.20	2.10	6.50	3.70
4	4.20	1.60	2.70	7.20	1.40
5	1.70	7.10	1.60	0.90	1.80
6	4.70	5.50	1.60	3.90	4.00
7	6.20	2.00	1.20	0.90	1.40
8	1.40	2.70	3.80	4.60	3.80
9	1.10	4.30	9.10	3.10	2.70
10	5.30	4.10	9.80	2.90	2.70
11	3.20	2.90	4.10	5.60	0.80
12	2.40	4.30	6.70	1.90	4.80
13	8.80	5.30	6.60	1.00	4.50
14	3.70	3.60	2.00	2.70	5.90
15	1.00	1.90	6.50	3.30	4.70
16	7.00	4.00	4.90	4.40	4.70
17	5.50	7.10	2.10	0.90	2.80
18	1.80	5.60	2.20	1.70	2.10
19	2.60	3.70	4.80	1.40	5.80
20	3.60	0.80	5.10	4.70	6.30

a. Construct an R-chart from these data.

d. Construct an $\overline{x}$ -chart from these data.

DDXL Output

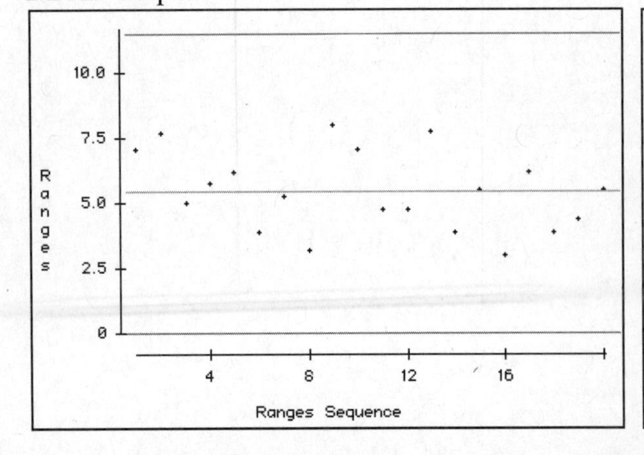

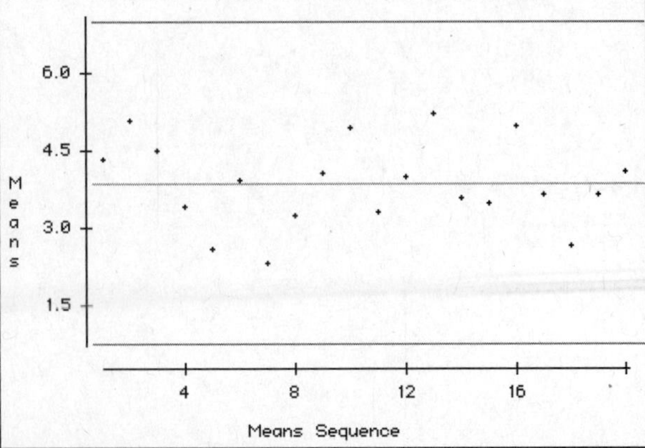

12.79 A company called CRW runs credit checks for a large number of banks and insurance companies. Credit history information is typed into computer files by trained administrative assistants. The company is interested in monitoring the proportion of credit histories that contain one or more data entry errors. Based on her experience with the data entry operation, the director of the data processing unit believes that the proportion of histories with data entry errors is about 6%. CRW audited 150 randomly selected credit histories each day for 20 days. The sample data are presented below.

Sample	Sample Size	Histories with Errors
1	150	9
2	150	11
3	150	12
4	150	8
5	150	10
6	150	6
7	150	13
8	150	9
9	150	11
10	150	5
11	150	7
12	150	6
13	150	12
14	150	10
15	150	11
16	150	7
17	150	6
18	150	12
19	150	14
20	150	10

b. Construct a p-chart for the data entry process.

DDXL Output

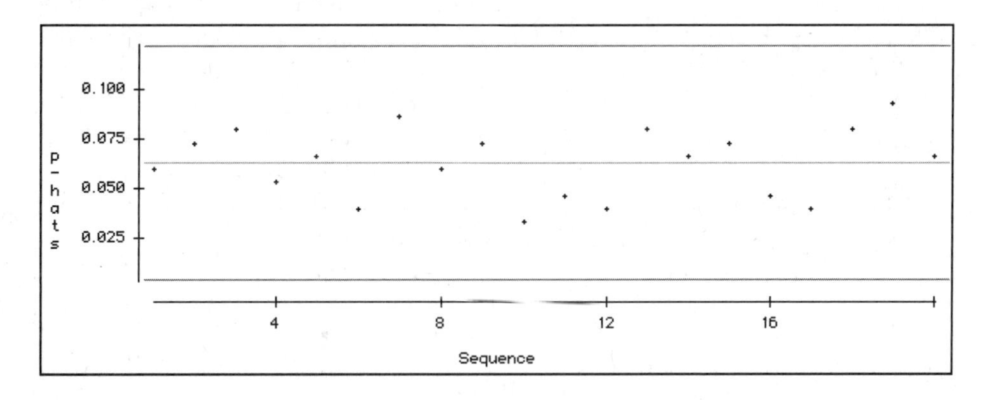

Chapter 13
Time Series: Descriptive Analyses, Models, and Forecasting

13.1 Introduction

Chapter 13 of the *Statistics for Business and Economics* text introduces the reader to the topic of Time Series Analysis. Descriptive analyses, time series modeling, and time series forecasting are the three main time series areas covered by the text.

Excel offers a variety of methods that enable the user to work with times series data. Many, like simple data manipulation, scatter plots, and regression analysis, have been encountered in one of the previous chapters of this manual. We reference these topics when we look at how they can be applied to times series data. Excel also offers times series tools that are not covered in the *Statistics for Business and Economics* text. Moving averages, Seasonal Indexes, and Cyclical Effects are topics offered within Excel but not covered in the text. We refer the reader to a more comprehensive text on time series analysis for information concerning these topics.

DDXL offers an easy technique for exponential smoothing that we will show below. The exponential smoothing technique that Excel offers differs from the exponential smoothing technique discussed in the book. We could, as an alternative, use simple formula manipulation of the time series data in Excel to get the desired exponentially smoothed values but choose instead to use the simple technique that DDXL offers.

There are several time series topics, however, that Excel is unable to provide assistance with. Section 13.5 of the text introduces the Holt-Winters forecasting model, an extension of the exponential smoothing topic covered in Section 13.2. Excel provides no data analysis tool to handle this more complicated smoothing model. In addition, the topic of measuring forecast accuracy (Section 13.6) has no Excel equivalent data analysis tool. And, finally, the Durbin-Watson test for autocorrelation is also not covered in Excel.

We will use the chapter examples that are given in the text to illustrate the model building and testing methods discussed above. The following examples from *Statistics for Business and Economics* are solved with Microsoft Excel® in the Chapter:

Excel Companion		**Statistics for Business and Economics**	
Exercise	**Page**	**Example**	**Excel File Name**
13.1	181	Example 13.1	HITECH
13.2	183	Example 13.2	Exercise 13.2
13.3	184	Example 13.3	Exercise 13.3
13.4	185	Example 13.4	SALES35

13.2 Descriptive Analyses: Index Numbers

Index numbers are the most common techniques for characterizing the change in a business or economic data series over time. These indexes can be constructed in a variety of manner. The text introduces the reader to the simple index, the simple composite index, and two different weighted composite indexes (Laspeyres and Paasche indexes). We examine how Excel can be used to generate these indexes in the examples that follow.

Exercise 13.1: We use Example 13.1 found in the *Statistics for Business and Economics* text.

One of the primary uses of index numbers is to characterize changes in stock prices over time. Stock market indexes have been constructed for many different types of companies and industries, and several composite indexes have been developed to characterize all stocks. These indexes are reported on a daily basis in the news media (e.g. Standard and Poor's 500 Stock Index and Dow Jones 65 Stock Index).

Consider the 2008 monthly closing prices (i.e., closing prices on the last day of each month) given in Table 13.1 for three high-technology company stocks listed on the New York Stock Exchange 5. To see how this type of stock fared, construct a simple composite index using January 2008 as the base period. Graph the index, and comment on its implications.

Solution:

Index values are found by taking the value of the series at some point in time and dividing by the value of the series during the base period and then multiplying this ratio by 100. Simple Composite indexes use totals from several different time series as the values in the index ratio. For this problem, we will use Excel to calculate the sum of the monthly closing prices of the listed stocks, and then use Excel to simply divide this sum by the sum found in the base year. The simple composite index will be found by multiplying this ratio by 100.

We use very basic data manipulation techniques within Excel to generate the desired indexes. We begin by **opening** the Excel file **HIGHTECH** by following the directions found in the preface of this manual. We will assume that the time series values are located in Columns D - F and Rows 2 - 13. The Column of totals should appear in Column G of the Excel worksheet. Note that Column B in the data sets has a Month label. This column will be used to create the desired scatter plot.

Table 13.1

YEAR	MONTH	TIME	IBM	INTEL	MICROSOFT
2008	JAN	1	107.11	21.10	32.60
2008	FEB	2	113.86	19.97	27.20
2008	MAR	3	115.14	21.18	28.38
2008	APR	4	120.70	22.26	28.52
2008	MAY	5	129.43	23.18	28.32
2008	JUN	6	118.53	21.48	27.51
2008	JUL	7	127.98	22.19	25.72
2008	AUG	8	121.73	22.87	27.29
2008	SEP	9	116.96	18.73	26.69
2008	OCT	10	92.97	16.03	22.33
2008	NOV	11	81.60	13.80	20.22
2008	DEC	12	84.16	14.66	19.44

Click on the cell located in Row 2 Column H. Enter **=(G2/$G$2)*100** in the cell. Excel should return the value 100 in the H2 cell. **Copy** the H2 cell to the cells located in Column H Rows 3 - 25 (e.g. H3 - H25). Compare the results returned by Excel (see Table 13.2) to those found in the text. (Note: It is important that the denominator of the formula listed above include the dollar signs as this tells Excel to use the G2 cell as the base level in all subsequent calculations).

Table 13.2

YEAR	MONTH	TIME	IBM	INTEL	MICROSOFT	TOTAL	INDEX
2008	JAN	1	107.11	21.10	32.60	160.81	100.00
2008	FEB	2	113.86	19.97	27.20	161.03	100.14
2008	MAR	3	115.14	21.18	28.38	164.70	102.42
2008	APR	4	120.70	22.26	28.52	171.48	106.64
2008	MAY	5	129.43	23.18	28.32	180.93	112.51
2008	JUN	6	118.53	21.48	27.51	167.52	104.17
2008	JUL	7	127.98	22.19	25.72	175.89	109.38
2008	AUG	8	121.73	22.87	27.29	171.89	106.89
2008	SEP	9	116.96	18.73	26.69	162.38	100.98
2008	OCT	10	92.97	16.03	22.33	131.33	81.67
2008	NOV	11	81.60	13.80	20.22	115.62	71.90
2008	DEC	12	84.16	14.66	19.44	118.26	73.54

The scatter plot can be drawn in Excel using the graphing techniques that were introduced in Chapter 2 of this manual. The scatterplot of the composite index is shown below.

Figure 13.1

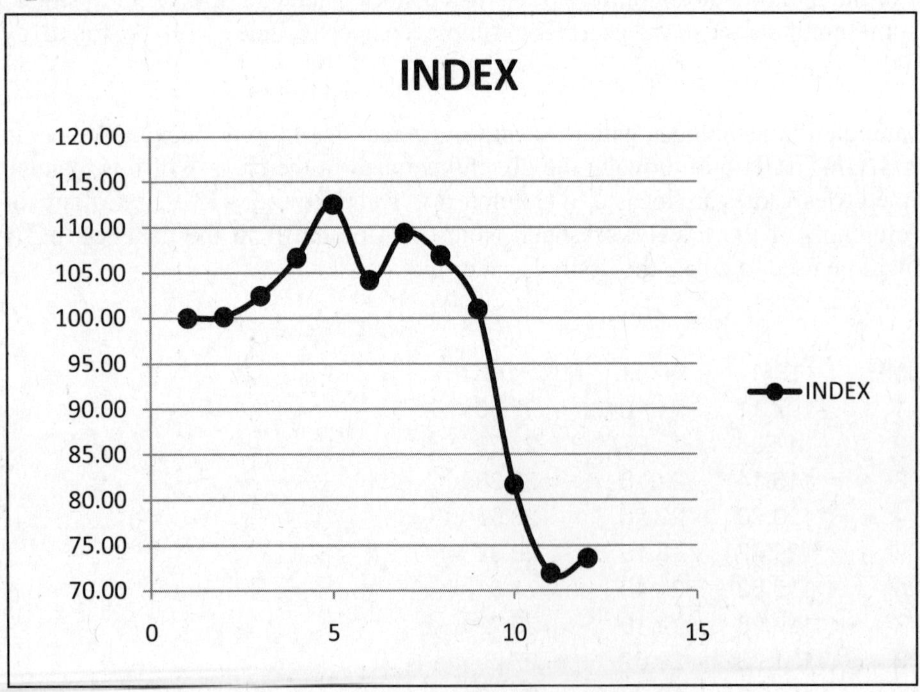

The simple composite index of Example 13.1 is found by summing the values of several times series and dividing by the sum from a base year. Each of the time series is given an equal weight in the simple composite index. Several different types of weighted composite indexes exist and the text discusses two such weighted composite indexes, Laspeyres and Paasche Indexes. We illustrate the Laspeyres Index below.

Exercise 13.2: We use Example 13.2 found in the *Statistics for Business and Economics* text.

The 2008 January and December prices for the three high technology company stocks are given in Table 13.3. Suppose that, in January 2008, an investor purchased the quantities shown in the table. [*Note*: Only two prices are used to simplify the example. The same methods can be applied to calculate the index for other months.] Calculate the Laspeyres index for the investor's portfolio of high-technology stocks using January 2008 as the base period.

Table 13.3

	IBM	Intel	Microsoft
Shares Purchased	500	100	1000
January 2008 Price	107.11	21.10	32.60
December 2008 Price	84.16	14.66	19.44

The first step in finding the Laspeyres indexes is to calculate the weighted price totals for each time period, using the January 2008 quantities as the weights. We begin by **opening** the Excel file **Exercise 13.2** by following the directions found in the preface of this manual. We will assume the weights are located in Column B - D, Row 2 in the worksheet and the time series values are located in columns B - D, Rows 3 - 4. **Click** on the cell located in Column E, Row 3. Enter **=$B$2*B3+$C$2*C3+$D$2*D3** in the cell. Make certain to use the "$" signs whenever you identify the cell locations of the weights. **Copy** the contents of Cell E3 to all the other cells in which weighted price totals are desired (e.g., Cell e4 in this example). Excel should return the weighted price totals shown below in Table 13.4.

The next step is to calculate the Laspeyres indexes using January 2008 as the base period. Click on the cell located in Column F, Row 3. Enter **=(E3/$E$3)*100** in the cell. **Copy** the contents of Cell F3 to all the other cells in which the Laspeyres indexes are desired (e.g., Cell F4 in this example). Excel should return the Laspeyres indexes shown below in Table 13.4.

Table 13.4

	IBM	Intel	Microsoft	Weighted Total	Laspeyres Index
Shares Purchased	500	100	1000		
January 2008 Price	107.11	21.1	32.6	88,265.00	100.00
December 2008 Price	84.16	14.66	19.44	62,986.00	71.36

Compare the results returned by Excel to those found in of the text.

As we have seen in the last example, the Laspeyres Index uses the quantities of the base period as the weights for all of the average and index values that are calculated. In some instances, it is preferred to use the quantities from the current time period as the weights in the index calculations. One method of achieving these indexes is to use the Paasche index. An example of the Paasche Index follows.

Exercise 13.3: We use Example 13.3 found in the *Statistics for Business and Economics* text.

The 2008 January and December prices and volumes (actual quantities purchased) in millions of shares for the three high-technology company stocks are shown in Table 13.5. Calculate and interpret the Paasche index, using January 2008 as the base period.

Table 13.5

	IBM		Intel		Microsoft	
	Price	Volume	Price	Volume	Price	Volume
January	107.11	247	21.10	1,462	32.60	1,950
December	84.16	189	14.66	1,332	19.44	1,549

Solution

The first step in finding the Paasche indexes in to calculate the weighted price totals for each time period, using the current time period quantities as the weights. We begin by **opening** the Excel file **Exercise 13.3** by following the directions found in the preface of this manual. **Click** on the specified cell in the worksheet and enter:

For Q_{Jan04}, P_{Jan04}: **=B4*C4+D4*E4+F4*G4** in cell A6
For Q_{Dec05}, P_{Dec05}: **=B3*C4+D3*E4+F3*G4** in cell A7

For the Paasche index for December, 2005, enter **=(A6/A7)*100** in cell A9. Excel should return the Paasche index shown below in Table 13.6. Compare the results returned by Excel to those found in the text.

Table 13.6

Q_{Jan}, P_{Jan}	65,545.92
Q_{Dec}, P_{Dec}	98,846.39
Paasche Index	66.31

Using indexes is just one method of describing time series data. We turn our attention now towards the use of exponential smoothing as a method to describe times series data.

13.3 Exponential Smoothing

A second method of describing time series data involves averaging past and current data together. The goal of the averaging is to reduce the volatility that is inherent to any time series. Exponential smoothing is one type of averaging method that allows the user to select the amount of weight given to the past and to the present data. This weight, known as the smoothing constant, is selected to be a number between 0 and 1. The larger the value of the smoothing constant, the more weight is given to the most current data value from the time series. We demonstrate with the following example.

Exercise 13.4: We use Example 13.4 found in the *Statistics for Business and Economics* text.

Problem: Annual sales date (recorded in thousands of dollars) for a firm's first 35 years of operation are provided in Table 3.7. Create an exponentially smoothed series for the sales time series using w=.7 and plot both series.

Table 13.7

T	SALES	T	SALES	T	SALES	T	SALES
1	5.8	11	35.5	21	86.2	31	136.3
2	4	12	53.5	22	89.9	32	146.8
3	5.5	13	48.4	23	89.2	33	150.1
4	15.6	14	61.6	24	99.1	34	151.4
5	25.1	15	65.6	25	100.3	35	150.9
6	20.3	16	71.4	26	111.7		
7	31.4	17	83.4	27	102.2		
8	48	18	93.6	28	115.5		
9	46.1	19	103.2	29	119.2		
10	35.9	20	85.4	30	125.2		

Solution:

We solve Exercise 13.4 utilizing the **Exponential Smoothing** function presented in the DDXL **Exponential Smoothing** menu (see Figure 13.2). **Open** the Data File **SALES35** by following the directions found in the preface of this manual. If done correctly, the data should appear in a workbook similar to that shown below in Figure 13.3. Use the mouse to select the data shown in the workbook.

Figure 13.2

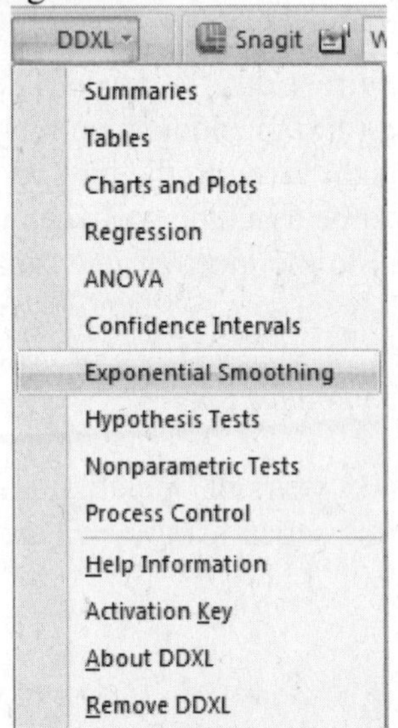

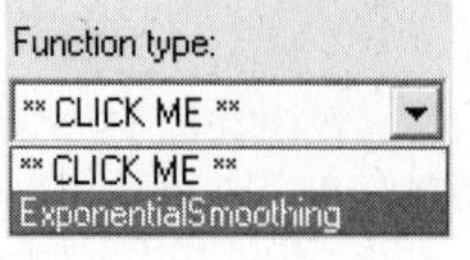

Figure 13.3

	A	B
1	T	SALES
2	1	5.8
3	2	4
4	3	5.5
5	4	15.6
6	5	25.1
7	6	20.3
8	7	31.4
9	8	48

Click on the **DDXL Add-In** menu. Click on the **Exponential Smoothing** option and click on the in the Function Type box to access the different procedures available to select. Highlight the **Exponential Smoothing** option to access the Exponential Smoothing Dialog menu shown in Figure 13.4. Highlight the **SALES** variable found in the **Names and Columns Box** and click on the to the right of the **Data Variable** box. Click **OK**. DDXL gives us the screen shown below in Figure 13.5.

Figure 13.4

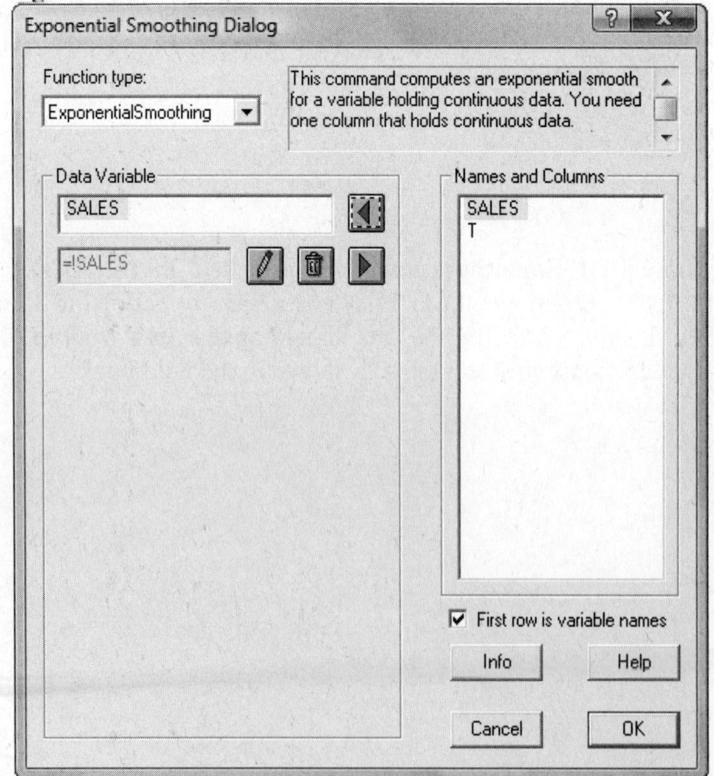

Figure 13.5

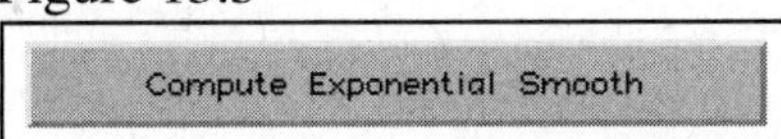

We click on the Compute Exponential Smooth button to create the exponentially smoothed series. DDXL automatically creates an exponentially smoothed series using a smoothing constant of *w*=.1. To change the smoothing constant, we click on the arrow to the right of the smoothing constant in the **Exponential Smoother Controls** box (see Figure 13.6). We select the **Set Value** option and type in the value we desire, *w*=.7, in this problem (see Figure 13.7). Click on Re-compute to generate the smoothed values (see Table 13.8) and a plot of the series (see Figure 13.8)..

Figure 13.6

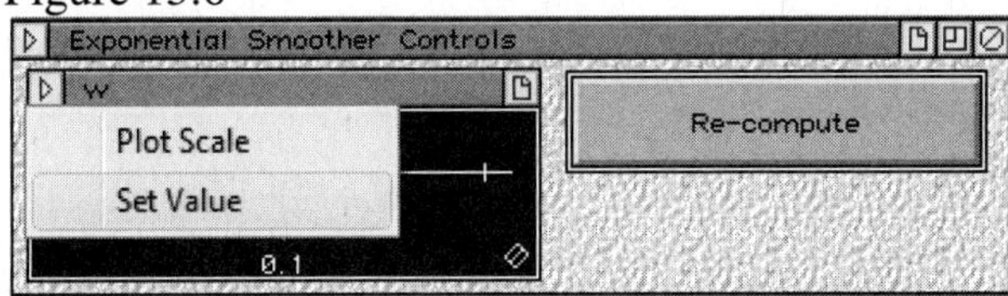

Figure 13.7

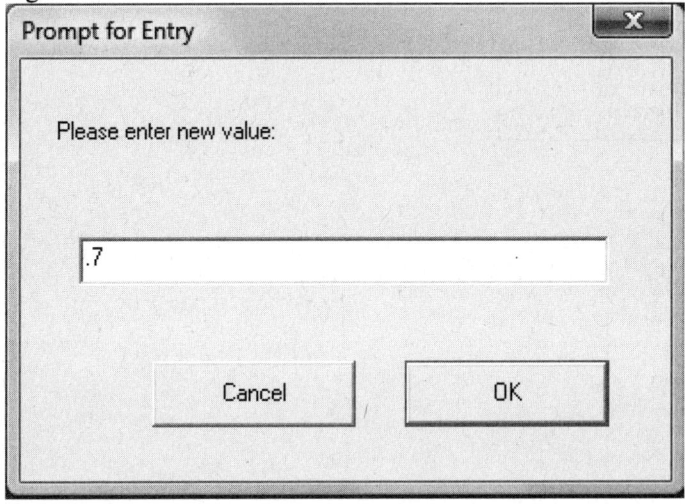

Table 13.8

Time Period	Actual Values	Smooth Values	Forecast Values
1	5.8	5.8	•
2	4	4.54	5.8
3	5.5	5.212	4.54
4	15.6	12.4836	5.212
5	25.1	21.31508	12.4836
6	20.3	20.604524	21.31508
7	31.4	28.161357	20.604524
8	48	42.048407	28.161357
9	46.1	44.884522	42.048407
10	35.9	38.595357	44.884522
11	35.5	36.428607	38.595357
12	53.5	48.378582	36.428607
13	48.4	48.393575	48.378582
14	61.6	57.638072	48.393575
15	65.6	63.211422	57.638072
16	71.4	68.943427	63.211422
17	83.4	79.063028	68.943427
18	93.6	89.238908	79.063028
19	103.2	99.011673	89.238908
20	85.4	89.483502	99.011673
21	86.2	87.185051	89.483502
22	89.9	89.085515	87.185051
23	89.2	89.165655	89.085515
24	99.1	96.119696	89.165655
25	100.3	99.045909	96.119696
26	111.7	107.90377	99.045909
27	102.2	103.91113	107.90377
28	115.5	112.02334	103.91113
29	119.2	117.047	112.02334
30	125.2	122.7541	117.047
31	136.3	132.23623	122.7541
32	146.8	142.43087	132.23623
33	150.1	147.79926	142.43087
34	151.4	150.31978	147.79926
35	150.9	150.72593	150.31978
36	•	•	150.72593

Figure 13.8

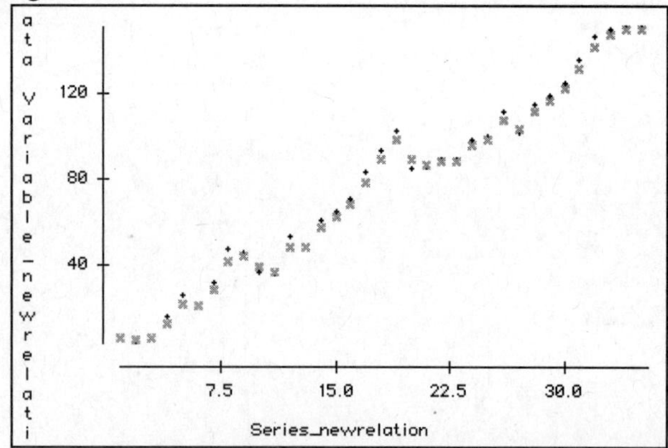

A plot of both the original times series the exponentially smoothed series is shown above in Figure 13.8. Compare this chart to the one shown in the text. Note that any value of a smoothing constant can be used by changing the appropriate values in the formula used to calculate the exponentially smoothed values of the series. The exponentially smoothed values will change depending whether the constant places more weight on the current value of the past values of the time series.

13.4 Using Regression to Model Time Series Data

Sections 13.7 and 13.8 introduce the reader to using regression models to model the linear trend and seasonal variation in time series data. The linear trend component can be modeled by using a measure of the time period as a quantitative variable in the regression. For example, the model $E(Y_t) = \beta_0 + \beta_1 t$ can be used to forecast the value of a time series at time period t. This model would assume that the time series values increase linearly over time.

While the linear model discussed above works in some applications, many times series data are affected by some sort of cyclical, or seasonal, influences. This cyclical variation can be modeled in regression using the qualitative variables discussed in Chapter 11. The seasonal component must be identified and explained using qualitative indicator variables. For example, the model

$$E(Y_t) = \beta_0 + \beta_1 t + \beta_2 Q_1 + \beta_3 Q_2 + \beta_4 Q_3$$

could be used to include both a linear trend (modeled with the quantitative time period variable, t) and a seasonal component (modeled with the three indicator variables Q_1, Q_2, and Q_3). The Q's in this model would be appropriate if the time series was influenced by some quarterly effect. For a monthly effect, the model would need to include eleven indicator variables.

We will not fit regression models to the time series data of Chapter 13. We remind the user that Excel requires the data set to include all of the independent variables to be included in the regression model. The variables must be in adjacent columns of the Excel worksheet. We refer the user to Chapters 10 and 11 of this manual to review how to fit regression models within Excel. The regression models fit will yield estimates to the values of the time series data of Chapter 13.

Technology Lab

The following exercise from the *Statistics for Business and Economics* text is given for you to practice the procedure covered in the text that is available within DDXL. Included with the exercise is the DDXL output that was generated to solve the problem.

13.55/13.56 Insured social security workers. Workers insured under the Social Security program are categorized as fully and permanently insured, fully but not permanently insured, or insured in the event of disability. The number of workers (in millions) in each insured category from 2000 to 2008 are provided in the accompanying table.

a. Compute a simple composite index for the number of workers in the three insured categories using 2000 as the base period.

b. Compute the exponentially smoothed series for the annual number of workers who are fully and permanently insured for Social Security. Use a smoothing constant of w=.5.

DDXL Output part a.

Year	Fully, permanent	Fully, not permanent	Event of disability	Total	Index
2000	140.9	44.9	139.5	325.3	100.00
2001	142.9	45.2	141.7	329.8	101.38
2002	144.9	45.3	143.5	333.7	102.58
2003	147.0	45.0	144.9	336.9	103.57
2004	149.0	44.8	146.2	340.0	104.52
2005	151.1	44.7	147.7	343.5	105.59
2006	153.3	45.1	150.1	348.5	107.13
2007	155.4	45.6	152.3	353.3	108.61
2008	157.4	46.0	154.5	357.9	110.02

DDXL Output part b.

Time Period	Actual Values	Smooth Values	Forecast Values
1	140.9	140.9	•
2	142.9	141.9	140.9
3	144.9	143.4	141.9
4	147	145.2	143.4
5	149	147.1	145.2
6	151.1	149.1	147.1
7	153.3	151.2	149.1
8	155.4	153.3	151.2
9	157.4	155.35	153.3
10	•	•	155.35

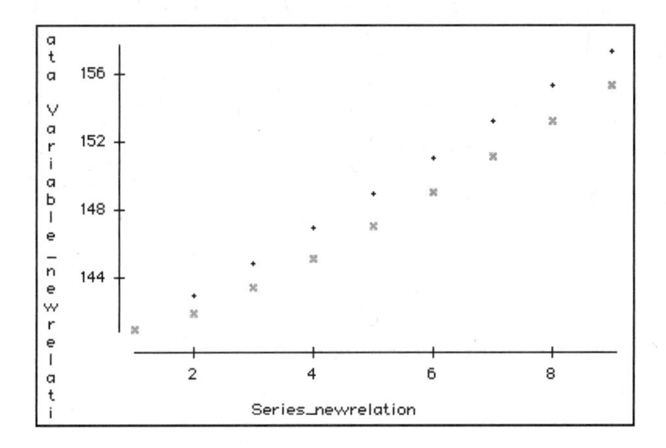

Chapter 14
Nonparametric Statistics

14.1 Introduction

Chapter 14 introduces the reader to the topic of nonparametric statistics and gives the reader several different examples of methods to analyze data using the nonparametric techniques. All of these techniques are available within the DDXL software program. The following examples from *Statistics for Business and Economics* are solved using DDXL in this chapter:

Excel Companion Exercise	Page	Statistics for Business and Economics Example	Excel File Name
14.1	192	Example 14.1	CDfailure
14.2	195	Example 14.2	Costliving
14.3	198	Example 14.3	Safety
14.4	201	Example 14.4	Hospeds
14.5	203	Example 14.5	Reaction
14.6	206	Example 14.6	Spoilage

14.2 The Sign Test

The *Statistics for Business and Economics* text offers the Sign Test as the technique for testing a single population median. DDXL allows the user to perform this test of hypothesis when the data that has been collected represent a single random sample. We illustrate how to use the Sign Test below.

Exercise 14.1: We use Example 14.1 found in the *Statistics for Business and Economics* text.

A manufacturer of compact disk (CD) players has established that the median time to failure for its players is 5,250 hours of utilization. A sample of 20 CDs from a competitor is obtained, and they are continuously tested until each fails. The 20 failure times range from 5 hours (a "defective player") to 6,575 hours, and 24 of the 40 exceed 5,250 hours. Is there evidence that the median failure time of the competitor differs from 5,250 hours? Use $\alpha = .10$.

Solution:

Figure 14.1

	A
1	Failure Times
2	5
3	875
4	1,125
5	2,545
6	3,515
7	4,775
8	5,321
9	5,338
10	5,407
11	5,485

We solve Exercise 14.1 utilizing the **Sign Test** presented in the **Nonparametric Tests** menu within DDXL. **Open** the Data File **CDfailure** by following the directions found in the preface of this manual. If done correctly, the data should appear in a workbook similar to that shown below in Figure 14.1. Use the mouse to select the data shown in the workbook.

Click on the **DDXL Add-In** menu. Click on the **Nonparametric Tests** option to access the various procedures available (see Figure 14.2). Click on the ▾ in the Function Type box to access the different tests available to select. Highlight the **Sign Test** option to access the Nonparametric Test Dialog menu shown in Figure 14.3. Highlight the **Failure Times** variable found in the **Names and Columns Box**. Click on the ◀ to the right of the **Quantitative Variable** box to create the test of hypothesis desired. Click **OK**. DDXL now asks you to specify the test of hypothesis that you desire to test (see Figure 14.4).

Figure 14.2

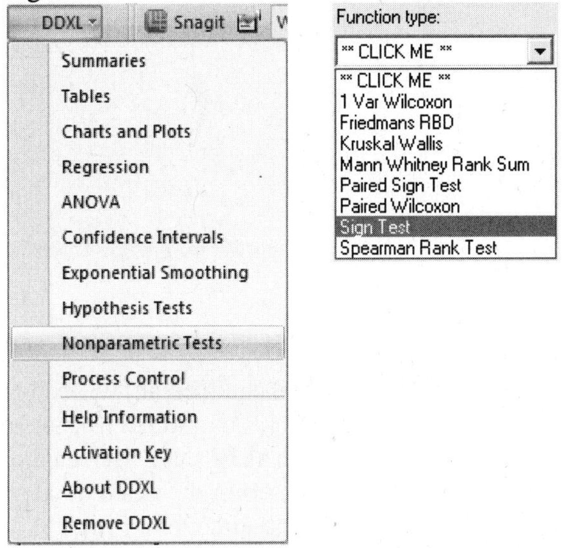

Figure 14.3

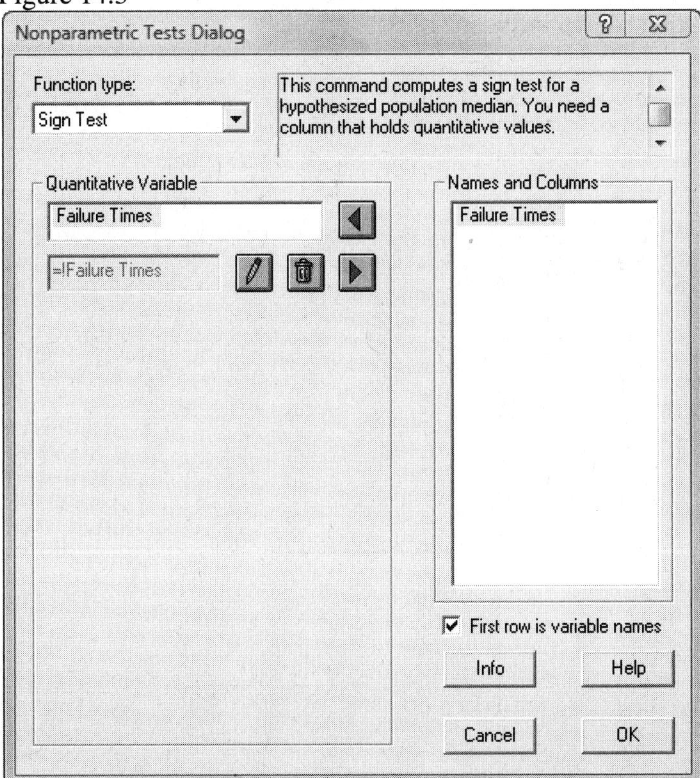

Figure 14.4

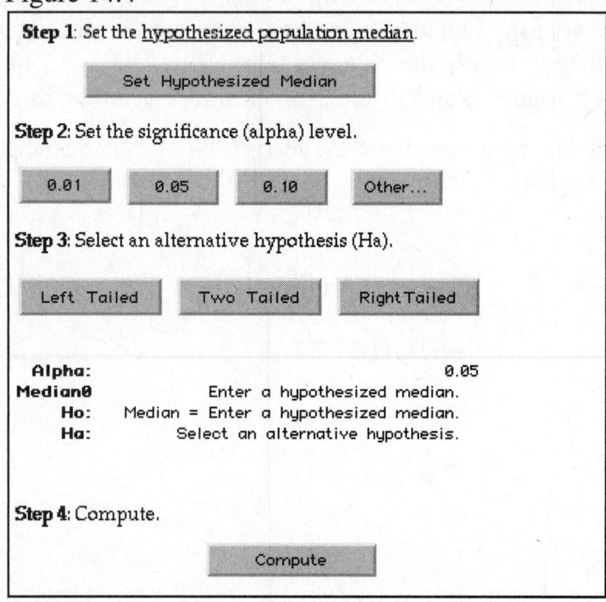

In **Step 1**, we click on the **Set Hypothesized Median** box to specify the value of the median we wish to test. In this problem, we specify a value of 5,250 (see Figure 14.5 below) and click **OK**. **Step 2** asks us to specify the alpha level that we desire to use for the test. Choices of **0.01**, **0.05**, or **0.10** can be selected or any other value of alpha can be entered by clicking on the **Other…** box and entering the desired value. For this problem, we click on the **0.10** box. **Step 3** requires the user to specify the direction of the test to be computed. The three alternative hypothesis choices (Left Tailed, Two Tailed, and Right Tailed) can be selected by clicking on the appropriate box. In this problem, we click on the **Two Tailed** box. Finally, in **Step 4**, we click on **Compute** to create the output shown in Figure 14.6.

Figure 14.5

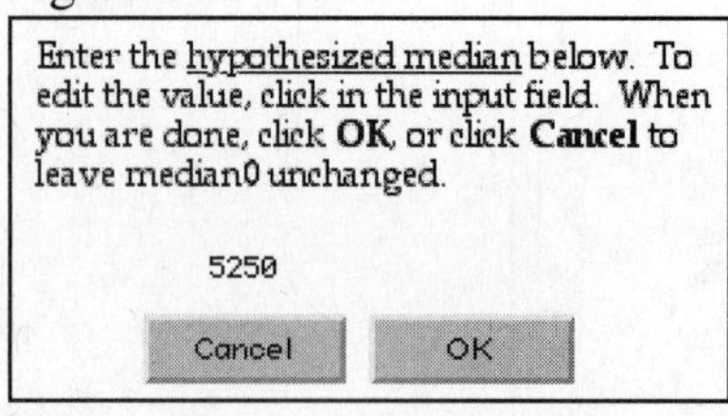

Figure 14.6

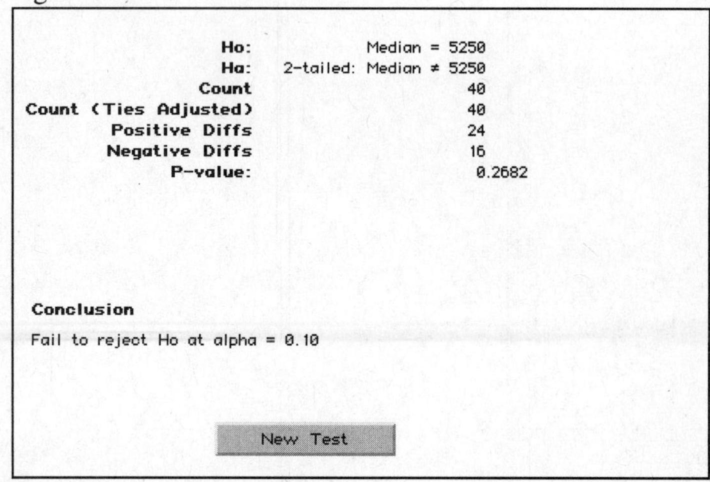

The DDXL printout gives us a p-value that can be used to make the appropriate conclusion. We see from the printout that the p-value for the test is p = .2682, and the appropriate conclusion for the test is fail to reject Ho when testing at alpha = .10. We compare these results with the ones given in the text to verify that we are conducting the test of hypothesis correctly.

14.3 The Mann-Whitney Test for Comparing Two Independent Samples

The *Statistics for Business and Economics* text offers the Wilcoxon Rank Sum technique for comparing two population means with independent samples. The procedure available in DDXL is the Mann-Whitney technique. As described in the text, the Mann-Whitney is equivalent to the Wilcoxon Rank Sum Test and gives identical p-values for the tests. DDXL allows the user to perform this test of hypothesis when the data has been collected using two random, independent samples. We illustrate how to use the Mann-Whitney Test below.

Exercise 14.2: We use Example 14.2 found in the *Statistics for Business and Economics* text.

Test the hypothesis that the government economists' prediction of next year's percentage change in cost of living tend to be lower than the university economists' – that is, *shifted to the left* of the probability distribution of university economists' predictions. Conduct the test using the data shown below in Table 14.1 and $\alpha = .05$.

Table 14.1

Government	University
3.1	4.4
4.8	5.8
2.3	3.9
5.6	8.7
0.0	6.3
2.9	10.5
	10.8

Solution:

We solve Exercise 14.2 utilizing the **Mann-Whitney Test** presented in the **Nonparametric Tests** menu within DDXL. **Open** the Data File **Costliving** by following the directions found in the preface of this manual. If done correctly, the data should appear in a workbook similar to that shown below in Figure 14.7. Use the mouse to select the data shown in the workbook.

Figure 14.7

	A	B
1		University
2	3.1	4.4
3	4.8	5.8
4	2.3	3.9
5	5.6	8.7
6	0.0	6.3
7	2.9	10.5
8		10.8

Click on the **DDXL Add-In** menu. Click on the **Nonparametric Tests** option to access the various procedures available (see Figure 14.8). Click on the ▾ in the Function Type box to access the different tests available to select. Highlight the **Mann Whitney Rank Sum** option to access the Nonparametric Test Dialog menu shown in Figure 14.9.

Figure 14.8

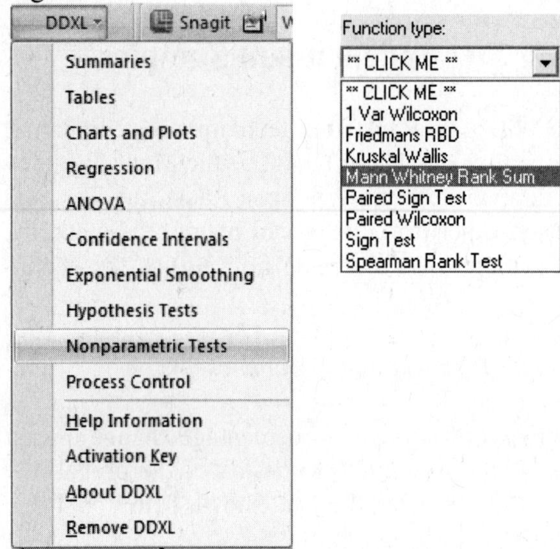

Figure 14.9

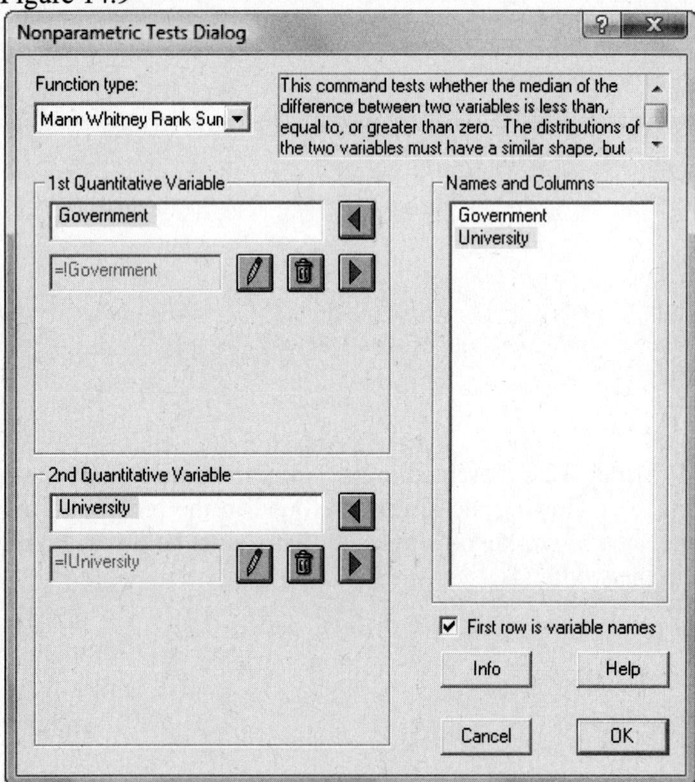

Highlight the **Government** variable found in the **Names and Columns Box**. Click on the ◀ to the right of the **1st Quantitative Variable** box. Next, highlight the **University** variable found in the **Names and**

Columns Box. Click on the to the right of the **2nd Quantitative Variable** box to create the test of hypothesis desired. Click **OK**. DDXL now asks you to specify the test of hypothesis that you desire to test (see Figure 14.10).

Figure 14.10

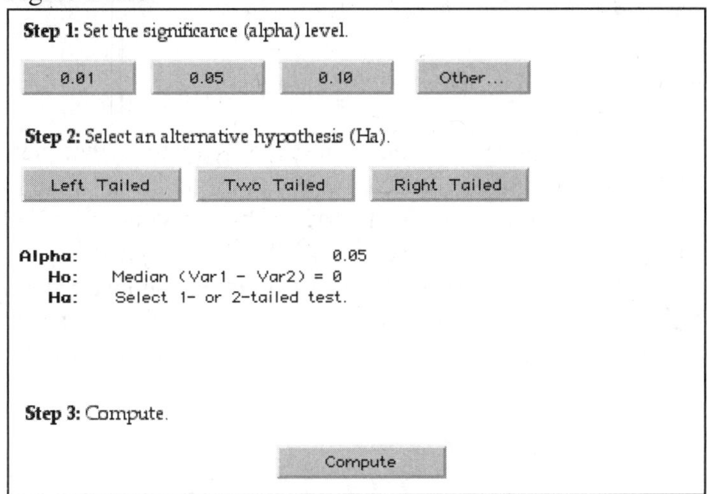

Step 1 asks us to specify the alpha level that we desire to use for the test. Choices of **0.01**, **0.05**, or **0.10** can be selected or any other value of alpha can be entered by clicking on the **Other…** box and entering the desired value. For this problem, we click on the **0.05** box. **Step 2** requires the user to specify the direction of the test to be computed. The three alternative hypothesis choices (Left Tailed, Two Tailed, and Right Tailed) can be selected by clicking on the appropriate box. In this problem, we click on the **Left Tailed** box. Finally, in **Step 3**, we click on **Compute** to create the output shown in Figure 14.11.

Figure 14.11

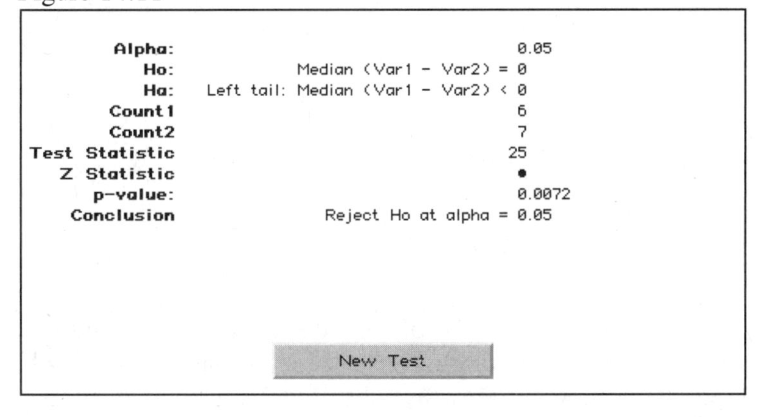

The DDXL printout gives us a p-value that can be used to make the appropriate conclusion. We see from the printout that the p-value for the test is p = .0072, and the appropriate conclusion for the test is reject Ho when testing at alpha = .05. We compare these results with the ones given in the text to verify that we are conducting the test of hypothesis correctly.

14.4 The Wilcoxon Signed Rank Test for Dependent Sampling

The *Statistics for Business and Economics* text offers the Wilcoxon Signed Rank Test comparing two population means with dependent samples. DDXL allows the user to perform this test of hypothesis when the data has been collected using a paired difference sampling design. We illustrate how to use the Wilcoxon Signed Rank Test below.

Exercise 14.3: We use Example 14.3 found in the *Statistics for Business and Economics* text.

Suppose the U.S. Consumer Product Safety Commission (CPSC) wants to test the hypothesis that New York City electrical contractors are more likely to install unsafe electrical outlets in urban homes than in suburban homes. A pair of homes, one urban and one suburban are both serviced by the same electrical contractor, is chosen for each of 10 randomly selected electrical contractors. A CPSC inspector assigns each of the 20 homes a safety rating between 1 and 10, with higher numbers implying safer electrical conditions. The results are shown in Table 14.2. Use the Wilcoxon signed rank test to determine whether the CPSC hypothesis is supported at the $\alpha = .05$ level.

Table 14.2

Contractor	Urban A	Suburban B
1	7	9
2	4	5
3	8	8
4	9	8
5	3	6
6	6	10
7	8	9
8	10	8
9	9	4
10	5	9

Solution:

Figure 14.12

	A	B	C
1	Contractor	Urban A	Suburban B
2	1	7	9
3	2	4	5
4	3	8	8
5	4	9	8
6	5	3	6
7	6	6	10
8	7	8	9
9	8	10	8
10	9	9	4
11	10	5	9

We solve Exercise 14.3 utilizing the **Paired Wilcoxon** test presented in the **Nonparametric Tests** menu within DDXL. **Open** the Data File **Safety** by following the directions found in the preface of this manual. If done correctly, the data should appear in a workbook similar to that shown below in Figure 14.12. Use the mouse to select the data shown in the workbook.

Click on the **DDXL Add-In** menu. Click on the **Nonparametric Tests** option to access the various procedures available (see Figure 14.13). Click on the ▾ in the Function Type box to access the different tests available to select. Highlight the **Paired Wilcoxon** option to access the Nonparametric Test Dialog menu shown in Figure 14.14.

Figure 14.13

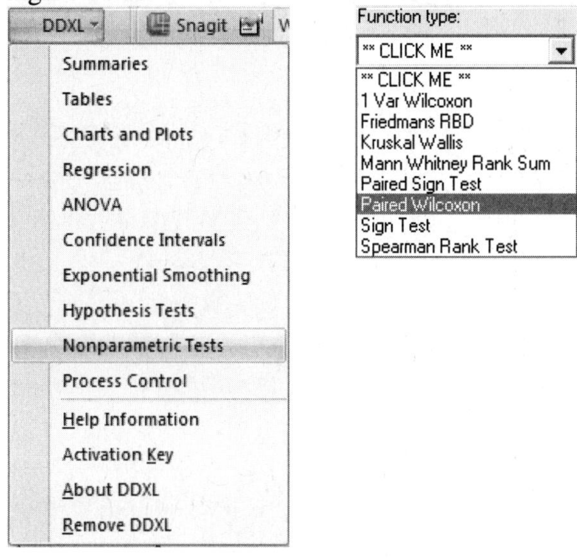

Figure 14.14

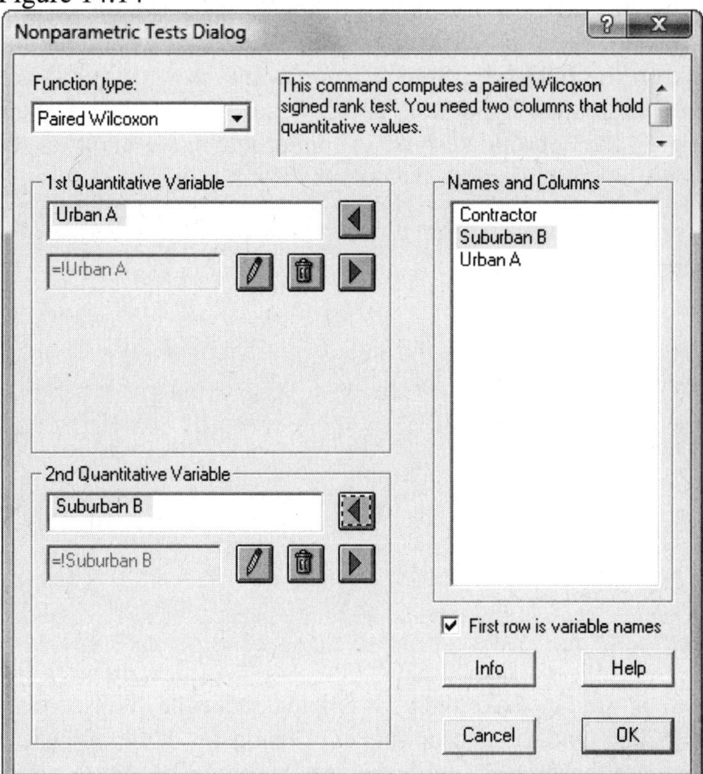

Highlight the **Urban A** variable found in the **Names and Columns Box**. Click on the to the right of the **1st Quantitative Variable** box. Next, highlight the **Suburban B** variable found in the **Names and Columns Box**. Click on the to the right of the **2nd Quantitative Variable** box to create the test of hypothesis desired. Click **OK**. DDXL now asks you to specify the test of hypothesis that you desire to test (see Figure 14.15).

Figure 14.15

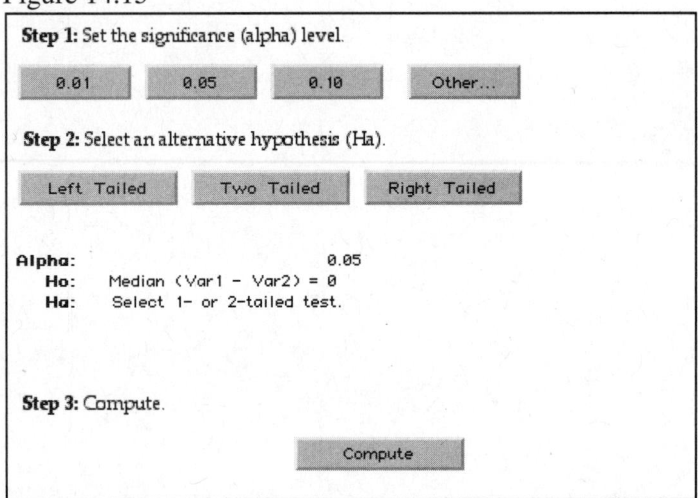

Step 1 asks us to specify the alpha level that we desire to use for the test. Choices of **0.01**, **0.05**, or **0.10** can be selected or any other value of alpha can be entered by clicking on the **Other...** box and entering the desired value. For this problem, we click on the **0.05** box. **Step 2** requires the user to specify the direction of the test to be computed. The three alternative hypothesis choices (Left Tailed, Two Tailed, and Right Tailed) can be selected by clicking on the appropriate box. In this problem, we click on the **Left Tailed** box. Finally, in **Step 3**, we click on **Compute** to create the output shown in Figure 14.16.

Figure 14.16

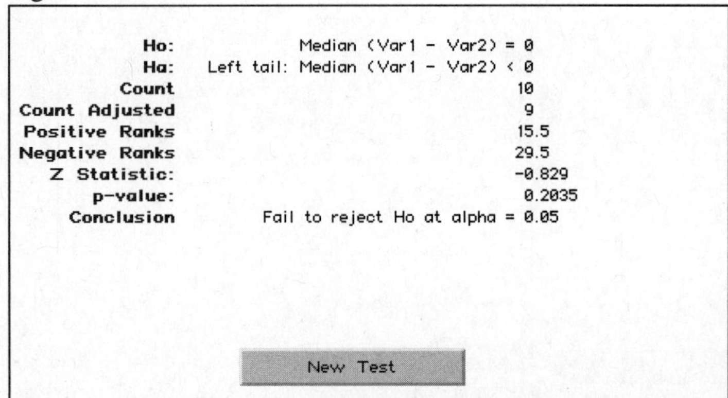

The DDXL printout gives us a p-value that can be used to make the appropriate conclusion. We see from the printout that the p-value for the test is p = .2035, and the appropriate conclusion for the test is fail to reject Ho when testing at alpha = .05. We compare these results with the ones given in the text to verify that we are conducting the test of hypothesis correctly.

Section 14.5: The Kruskal-Wallis H-Test for a Completely Randomized Design

The *Statistics for Business and Economics* text offers the Kruskal-Wallis H-Test for comparing three or means collected using a completely randomized sampling design. DDXL allows the user to perform this test of hypothesis utilizing the **Kruskal Wallis** option form the **Nonparametric Tests** menu. We illustrate how to use the Kruskal-Wallis Test below.

Exercise 14.4: We use Example 14.4 found in the *Statistics for Business and Economics* text.

Consider the data in Table 14.3. Recall that a health administrator wants to compare the unoccupied bed space of the three hospitals. Apply the Kruskal-Wallis H-test to the data. What conclusions can you draw? Test using α .05.

Table 14.3

Hospital 1	Hospital 2	Hospital 3
6	34	13
38	28	35
3	42	19
17	13	4
11	40	29
30	31	0
15	9	7
16	32	33
25	39	18
5	27	24

Solution:

Figure 14.17

	A	B
1	Beds	Hospital
2	6	1
3	38	1
4	3	1
5	17	1
6	11	1
7	30	1
8	15	1
9	16	1
10	25	1
11	5	1
12	34	2
13	28	2
14	42	2

We solve Exercise 14.4 utilizing the **Kruskal Wallis** test presented in the **Nonparametric Tests** menu within DDXL. **Open** the Data File **Hospbeds** by following the directions found in the preface of this manual. If done correctly, the data should appear in a workbook similar to that shown below in Figure 14.17. Use the mouse to select the data shown in the workbook.

Click on the **DDXL Add-In** menu. Click on the **Nonparametric Tests** option to access the various procedures available (see Figure 14.18). Click on the ▾ in the Function Type box to access the different tests available to select. Highlight the **Kruskal Wallis** option to access the Nonparametric Test Dialog menu shown in Figure 14.19.

Figure 14.18

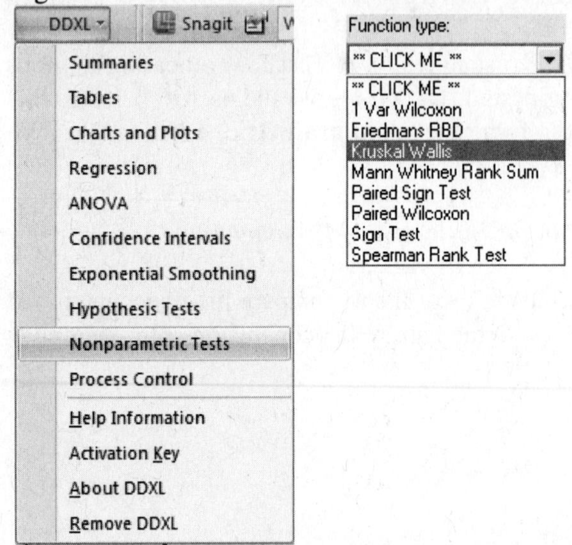

Figure 14.19

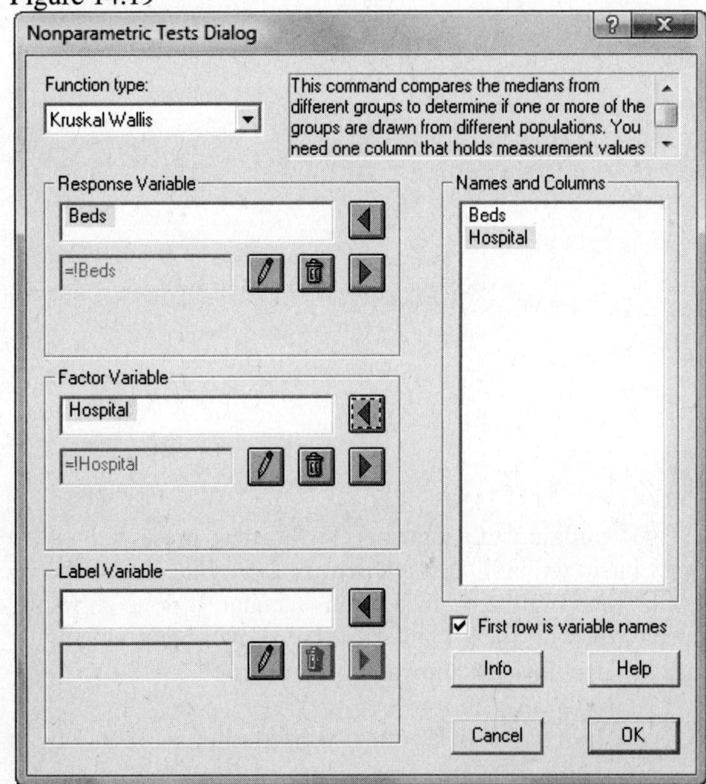

Highlight the **Beds** variable found in the **Names and Columns Box**. Click on the to the right of the **Response Variable** box. Next, highlight the **Hospital** variable found in the **Names and Columns Box**. Click on the to the right of the **Factor Variable** box to create the test of hypothesis desired. Click **OK** to generate the output shown in Figure 4.20.

Figure 14.20

T	6.097
p	0.0474
number of ties	1
T (corrected for ties)	6.099
p (corrected)	0.0474

Group	Count	Sum of Ranks	Mean Rank
1	10	120	12
2	10	210.5	21.05
3	10	134.5	13.45

The DDXL printout gives us a p-value that can be used to make the appropriate conclusion. We see from the printout that the p-value for the test is p = .0474, and the appropriate conclusion for the test is to reject Ho when testing at alpha = .05. We compare these results with the ones given in the text to verify that we are conducting the test of hypothesis correctly.

Section 14.6: The Friedman F_r-Test for a Randomized Block Design

The *Statistics for Business and Economics* text offers the Friedman F_r-Test for comparing three or means collected using a randomized block sampling design. DDXL allows the user to perform this test of hypothesis utilizing the **Friedman RBD** option form the **Nonparametric Tests** menu. We illustrate how to use the Friedman F_r-Test below.

Exercise 14.5: We use Example 14.5 found in the *Statistics for Business and Economics* text.

Consider the data in Table 14.4. Recall that a pharmaceutical firm wants to compare the reaction times of subjects under the influence of three different drugs that it produces. Apply the Friedman F_r-test to the data. What conclusions can you draw? Test using α .05.

Table 14.4

Subject	Drug A	Drug B	Drug C
1	1.21	1.48	1.56
2	1.63	1.85	2.01
3	1.42	2.06	1.70
4	2.43	1.98	2.64
5	1.16	1.27	1.48
6	1.94	2.44	2.81

Solution:

We solve Exercise 14.5 utilizing the **Friedman RBD** test presented in the **Nonparametric Tests** menu within DDXL. **Open** the Data File **Reaction** by following the directions found in the preface of this manual. If done correctly, the data should appear in a workbook similar to that shown below in Figure 14.21. Use the mouse to select the data shown in the workbook.

Figure 14.21

	A	B	C
1	Subject	Reaction Time	Drug Type
2	1	1.21	A
3	2	1.63	A
4	3	1.42	A
5	4	2.43	A
6	5	1.16	A
7	6	1.94	A
8	1	1.48	B
9	2	1.85	B
10	3	2.06	B

Click on the **DDXL Add-In** menu. Click on the **Nonparametric Tests** option to access the various procedures available (see Figure 14.22). Click on the in the Function Type box to access the different tests available to select. Highlight the **Friedman RBD** option to access the Nonparametric Test Dialog menu shown in Figure 14.23.

Figure 14.22

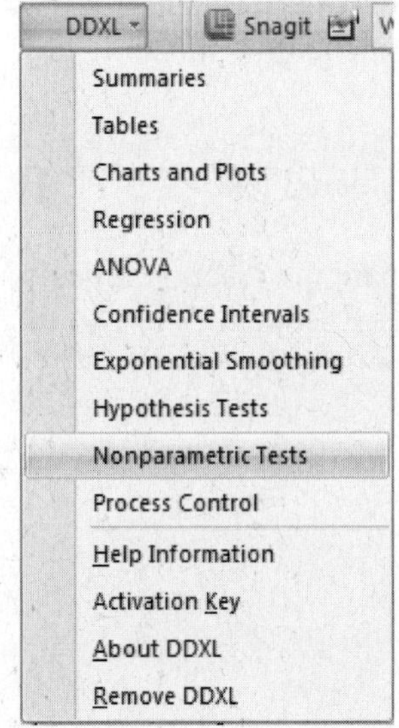

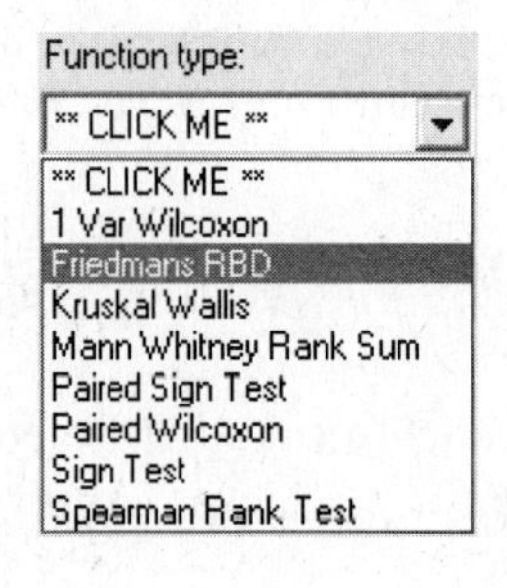

Highlight the **Reaction Time** variable found in the **Names and Columns Box**. Click on the to the right of the **Response** box. Next, highlight the **Drug Type** variable found in the **Names and Columns Box**. Click on the to the right of the **Factor Variable** box. Finally, highlight the **Subject** variable found in the **Names and Columns Box**. Click on the to the right of the **Block** box to create the test of hypothesis desired. Click **OK** to generate the output shown in Figure 4.24.

Figure 14.23

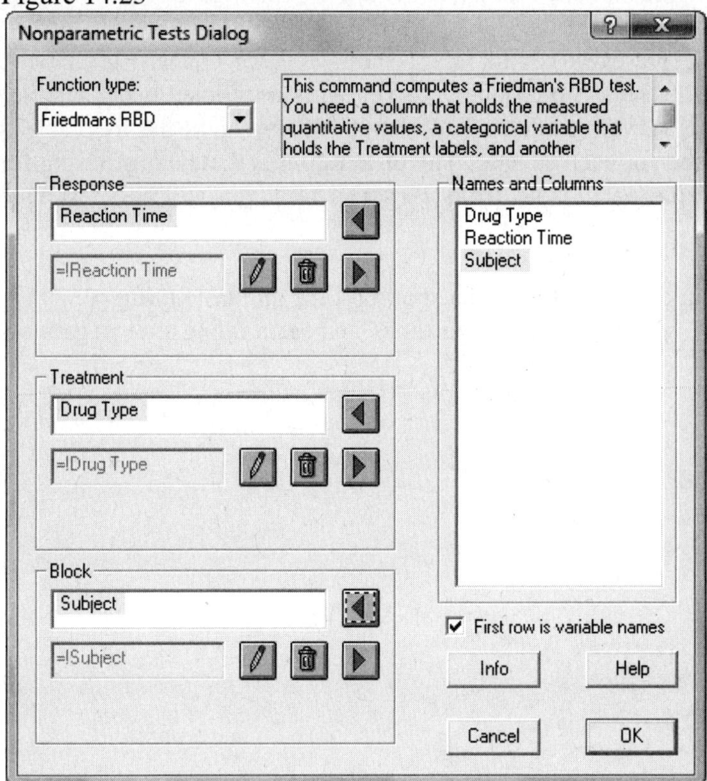

Figure 4.24

S	8.333
df	2
p-value	0.0155

Treatment Name	Response Median	Count
A	1.525	6
B	1.915	6
C	1.855	6

The DDXL printout gives us a p-value that can be used to make the appropriate conclusion. We see from the printout that the p-value for the test is p = .0155, and the appropriate conclusion for the test is to reject Ho when testing at alpha = .05. We compare these results with the ones given in the text to verify that we are conducting the test of hypothesis correctly.

Section 14.7: Spearman's Rank Correlation Coefficient

The *Statistics for Business and Economics* text offers the Spearman's Rank Correlation Coefficient as the method to test to determine if two variables are correlated. DDXL allows the user to perform this test of hypothesis utilizing the **Spearman Rank Test** form the **Nonparametric Tests** menu. We illustrate how to use the Friedman F_r-Test below.

Exercise 14.6: We use Example 14.6 found in the *Statistics for Business and Economics* text.

Manufacturers of perishable foods often use preservatives to retard spoilage. One concern is that too much preservative will change the flavor of the food. Suppose an experiment is conducted using samples of a food product with varying amounts of preservative added. Both length of time until the food shows signs of spoiling and a taste rating are recorded for each sample. The taste rating is the average rating for three tasters, each of whom rates each sample on a scale from 1 (good) to 5 (bad). Twelve sample measurements are shown in Table 14.5.

a. Calculate Spearman's rank correlation coefficient between spoiling time and taste rating.
b. Use a nonparametric test to find out whether the spoilage times and taste ratings are negatively correlated. Test using $\alpha = .05$.

Table 14.5

Sample	Days Until Spoilage	Taste Rating
1	30	4.3
2	47	3.6
3	26	4.5
4	94	2.8
5	67	3.3
6	83	2.7
7	36	4.2
8	77	3.9
9	43	3.6
10	109	2.2
11	56	3.1
12	70	2.9

Solution:

We solve Exercise 14.6 utilizing the **Spearman Rank Test** presented in the **Nonparametric Tests** menu within DDXL. **Open** the Data File **Spoilage** by following the directions found in the preface of this manual. If done correctly, the data should appear in a workbook similar to that shown below in Figure 14.25. Use the mouse to select the data shown in the workbook.

Figure 14.25

	A	B	C
1	Sample	Days Until Spoilage	Taste Rating
2	1	30	4.3
3	2	47	3.6
4	3	26	4.5
5	4	94	2.8
6	5	67	3.3
7	6	83	2.7
8	7	36	4.2

Click on the **DDXL Add-In** menu. Click on the **Nonparametric Tests** option to access the various procedures available (see Figure 14.26). Click on the ▾ in the Function Type box to access the different tests available to select. Highlight the **Spearman Rank Test** option to access the Nonparametric Test Dialog menu shown in Figure 14.27.

Figure 14.26

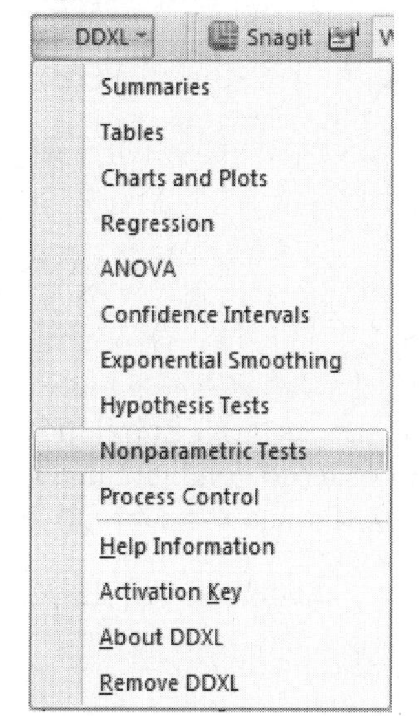

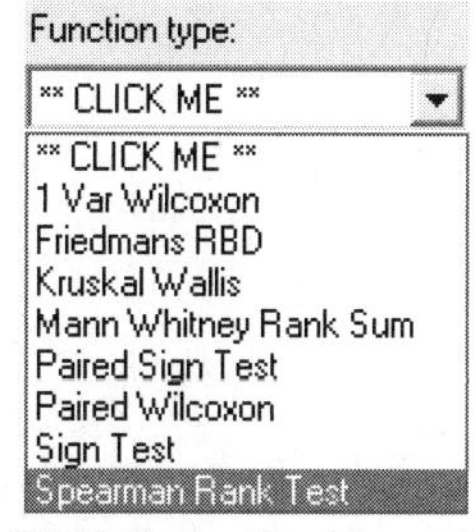

Highlight the **Days Until Spoilage** variable found in the **Names and Columns Box**. Click on the ◀ to the right of the **x-Axis Quantitative Variable** box. Next, highlight the **Taste Rating** variable found in the **Names and Columns Box**. Click on the ◀ to the right of the **y-Axis Quantitative Variable** box and click **OK** to generate the output shown in Figure 4.28.

Figure 14.27

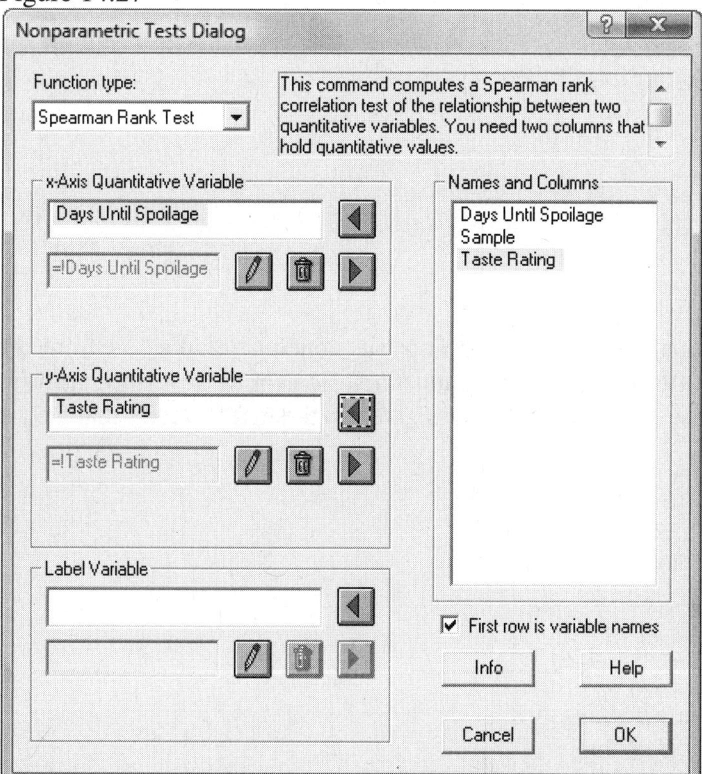

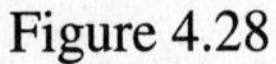
Figure 4.28

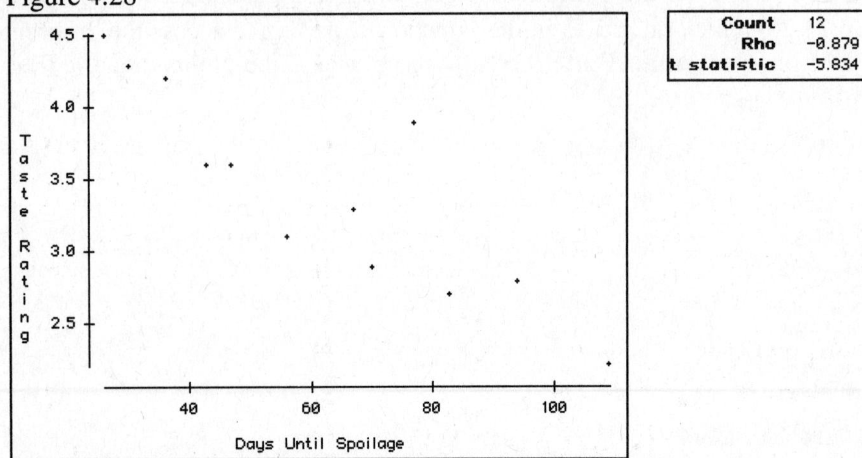

The DDXL printout gives both a plot of the two variables as well as the Spearman rank correlation coefficient. We see from the printout that the correlation coefficient is -0.879. In order to conduct the test of hypothesis, we select the correct test from the choices shown in Figure 4.29. To test for a negative correlation between the variables, we click on the Perform a Left Tailed Test box. The results are shown in Figure 14.30

Figure 14.29

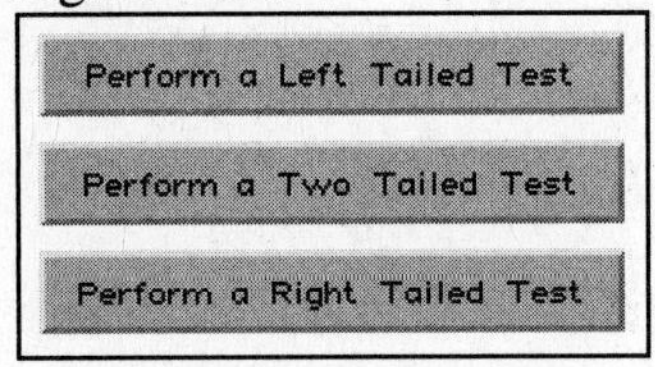

Figure 14.30

Ha	Left Tailed
t statistic	-5.834
p-value	0.0001

DDXL provide us a p-value that can be used to make the appropriate conclusion. We see from the printout that the p-value for the test is p = .0001, and the appropriate conclusion for the test is to reject Ho when testing at alpha = .05. We compare these results with the ones given in the text to verify that we are conducting the test of hypothesis correctly.

Technology Lab

The following exercises from the *Statistics for Business and Economics* text are given for you to practice the nonparametric procedures that are available within DDXL. Included with the exercises are the DDXL outputs that were generated to solve the problems.

14.88 An economist is interested in knowing whether property tax rates differ among three types of school districts – urban, suburban, and rural. A random sample of several districts of each type produced the data in the table (rate is in mills, where 1 mill = $1/1,000). Do the data indicate a difference in the level of property taxes among the three types of school districts? Use α = .05.

Urban	Suburban	Rural
4.3	5.9	5.1
5.2	6.7	4.8
6.2	7.6	3.9
5.6	4.9	6.2
3.8	5.2	4.2
5.8	6.8	4.3
4.7		

DDXL Output

T	5.85
p	0.0537
number of ties	3
T (corrected for ties)	5.865
p (corrected)	0.0533

Group	Count	Sum of Ranks	Mean Rank
Rural	6	41	6.833
Suburban	6	86.5	14.417
Urban	7	62.5	8.929

14.93 A hotel had a problem with people reserving rooms for a weekend and then not honoring their reservations (no-shows). As a result, the hotel developed a new reservation and deposit plan that it hoped would reduce the number of no-shows. One year after the policy was initiated; management evaluated its effect in comparison with the old policy. Compare the records given in the table on the number of no-shows for the 10 non-holiday weekends preceding the institution of the new policy and the 10 non-holiday weekends proceeding the evaluation time. Has the situation improved under the new policy? Test at α =.05.

Before	After
10	4
5	3
3	8
6	5
7	6
11	4
8	2
9	5
6	7
5	1

DDXL Output

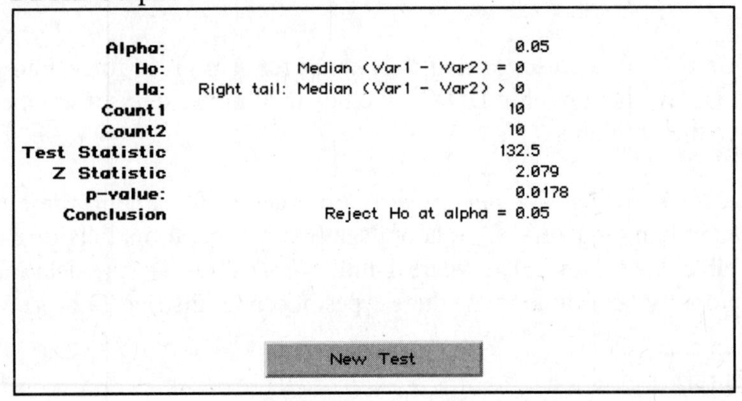

Alpha:	0.05
Ho:	Median (Var1 - Var2) = 0
Ha:	Right tail: Median (Var1 - Var2) > 0
Count1	10
Count2	10
Test Statistic	132.5
Z Statistic	2.079
p-value:	0.0178
Conclusion	Reject Ho at alpha = 0.05

New Test

14.95 A union wants to determine the preferences of its members before negotiating with management. Ten union members are randomly selected, and an extensive questionnaire is completed by each member. The responses to the various aspects of the questionnaire will enable the union to rank in order of importance the items to be negotiated. The rankings are shown in the table below. Conduct a nonparametric test to determine whether evidence exists that the probability distributions of ratings differ for at least two of the four negotiable items. Use $\alpha = .05$.

Person	More Pay	Job Stability	Fringe Benefits	Shorter Hours
1	2	1	3	4
2	1	2	3	4
3	4	3	2	1
4	1	4	2	3
5	1	2	3	4
6	1	3	4	2
7	2.5	1	2.5	4
8	3	1	4	2
9	1.5	1.5	3	4
10	2	3	1	4

DDXL Output

S	6.21
df	3
p-value	0.1018

Treatment Name	Response Median	Count
Fringe Benefits	3	10
Job Stability	2	10
More Pay	1.75	10
Shorter Hours	4	10

14.97 Agent Orange – the code name for a herbicide developed for the U.S. Armed Forces in the 1960's – was found to be extremely contaminated with TCDD, or dioxin. During the Vietnam War, and estimated 19 million gallons of Agent Orange was used to destroy the dense plant and tree cover of the Asian jungle. As a result of this exposure, many Vietnam veterans have dangerously high levels of TCDD in their blood and adipose (fatty) tissue. A study published in *Chemosphere* (Vol. 20, 1990) reported on the TCDD levels of 20 Massachusetts Vietnam vets who were possibly exposed to Agent Orange. The TCDD amounts (measured in parts per trillion) in both plasma and fat tissue of the 20 vets are listed in the accompanying table.

a. Medical researchers consider a TCDD level of 3 parts per trillion (ppt) to be dangerously high. Do the data provide evidence (at $\alpha = .05$) to indicate that the median level of TCDD in the fat tissue of Vietnam vets exceeds 3 ppt?

b. Repeat part a for plasma.

c. Medical researchers also are interested in comparing the TCDD levels in fat tissue and plasma for Vietnam veterans. Specifically, they want to determine if the distribution of TCDD levels in fat is shifted above or below the distribution of TCDD levels in plasma. Conduct this analysis (at $\alpha = .05$) and make the appropriate inference.

d. Find the rank correlation between the TCDD level in fat tissue and the TCDD level in plasma. Is there sufficient evidence (at $\alpha = .05$) of a positive association between the two TCDD measures?

Vet	Fat	Plasma
1	4.9	2.5
2	6.9	3.5
3	10.0	6.8
4	4.4	4.7
5	4.6	4.6
6	1.1	1.8
7	2.3	2.5
8	5.9	3.1
9	7.0	3.1
10	5.5	3.0
11	7.0	6.9
12	1.4	1.6
13	11.0	20.0
14	2.5	4.1
15	4.4	2.1
16	4.2	1.8
17	41.0	36.0
18	2.9	3.3
19	7.7	7.2
20	2.5	2.0

DDXL Output 14.97 a

Ho:	Median = 3
Ha:	Upper tail: Median > 3
Count	20
Count (Ties Adjusted)	20
Positive Diffs	14
Negative Diffs	6
P-value:	0.0577

Conclusion

Fail to reject Ho at alpha = 0.05

New Test

DDXL Output 14.97 b

Ho:	Median = 3
Ha:	Upper tail: Median > 3
Count	20
Count (Ties Adjusted)	19
Positive Diffs	12
Negative Diffs	7
P-value:	0.1796

Conclusion

Fail to reject Ho at alpha = 0.05

New Test

DDXL Output 14.97 c

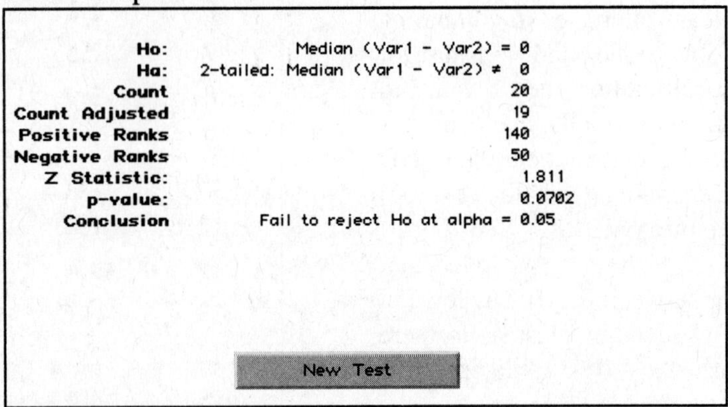

Ho:	Median (Var1 - Var2) = 0
Ha:	2-tailed: Median (Var1 - Var2) ≠ 0
Count	20
Count Adjusted	19
Positive Ranks	140
Negative Ranks	50
Z Statistic:	1.811
p-value:	0.0702
Conclusion	Fail to reject Ho at alpha = 0.05

New Test

DDXL Output 14.97 d

Count	20
Rho	0.777
t statistic	5.232

Ha	Right Tailed
t statistic	5.232
p-value	0